Key Concepts

STEM

SCIENCE TECHNOLOGY ENGINEERING MATH

ENGINEERING AND TECHNOLOGY

MILITARY AND SECURITY

General Editor: Anne O'Daly

BROWN BEAR BOOKS

Published by Brown Bear Books Ltd
4877 N. Circulo Bujia, Tucson, AZ 85718
USA
and
Studio G14, Regent Studios, 1 Thane Villas,
London N7 7PH, UK

Cataloging-in-Publication data available upon request.

ISBN 978-1-83572-064-6 (ALB)
ISBN 978-1-83572-070-7 (paperback)
ISBN 978-1-83572-076-9 (ebook)

Children's Publisher: Anne O'Daly
Design Manager: Keith Davis
Picture Manager: Sophie Mortimer
Consultant: Chris Woodford

Manufactured in the United States of America
CPSIA compliance information: Batch#AG/5664

Picture Credits
Front Cover: Shutterstock: Mike Mareen. **Interior: iStock:** 4br, 7bl, 10t, baona 58t, enviromantic 55b, Maksims Grigorjevs 54, Hemera 16, 20–21, 26t, 33, 50, 28b, 34–35, malerapaso 55t, MR1805 41, Photos.com 13t, 14–15, 20b, 22b, 23, 42, 44, pidjoe 56b, Chainarong Prasertthai 56t, RASimon 24, SpiffyJ 59t, turk stock photographer 58b; **Library of Congress:** 46; **Public Domain:** Bibliotetheque Nationale de France 6–7t, 60, James F. Gibson 18, Los Angeles County Museum of Art 4, Ben Mills 22t, Woolford 26b, 61t; **Robert Hunt Library:** 21, 28t; **Shutterstock:** Algol 8t; B-1972 6b, Denis Barbulat 8b, 11br, Belish 19, Bibiphoto 27t, Bonita Cheshier 53t, Ensuper 49t, ES3D Studios 49b, Everett Historical 30, 53b, Gerasimov foto 35b, GLYPHstock 27b, Gorodenkoff 39t, jules2000 39b, Jumpingjack 25t, Szynmon Kaczmarczyk 29, Yu Lan 12, David Lech 13b, Lenscap Photography 32t, Peter Lorimer 15, Mar.K 48, Militarist 25b, New Punisher 5b, Sergey Nivens 1, 59b, Sailful Nizam 45, Regien Paassen 4–5, Peyker 12b, PressLab 36b, ShutterStockStudio 42-43, John Silver 61b, TFoxFoto 32b, Phil Trott 52, Daniel Vega-Ayala 47, John Wollwerth 37, yuri 36t; **United States Government:** Department of Defense/John F Williams/U.S. Navy 40. NARA/Department of Defense 34b.

STEM

Key Concepts in STEM: Engineering and Technology describes the scientific principles and feats of engineering that have shaped the world we live in. Since earliest times, people have created tools and machines to provide shelter, warmth, security, and protection from disease. This book, *Military and Security*, looks at the technology of weaponry from the earliest days of clubs and daggers through to drones and the use of artificial intelligence in warfare. It also charts the evolution of security, from early locks and keys to biometric security devices. Explanatory diagrams and informative photographs explain how STEM (science, technology, engineering, and math) has transformed our lives and continues to influence our world today.

Contents

WARFARE BEFORE 1850

People have been fighting wars since ancient times. In any battle, the right weapons can make the difference between winning and losing.

The first battles were probably fought with hunting tools such as spears and stone daggers. The earliest weapon to be designed specifically for the purpose of warfare is thought to be the club, or mace. The earliest maces, made at the beginning of the Bronze Age (10,000 BCE), consisted of a large stone attached to a wooden handle, and were designed to crush the skull of an enemy.

THE SAMURAI SWORD

The first iron swords were made about 3,200 years ago. Swords have taken many forms since then, but many believe that the art of the swordsmith reached its high point in the *katana*, the weapon of the Japanese samurai warriors. Before it was put in a furnace, the *katana's* blade, apart from the cutting edge, was coated in clay. This unique process allowed the sword to combine extreme flexibility with a razor-sharp blade.

A katana *in its ornate sheath (left), and the smaller* wakizashi *sword and sheath (far left).*

In the third millennium BCE advances in metalworking allowed some mace heads to be made from copper, making the mace a more effective weapon.

As offensive weapons became more advanced, so did defensive armor. By 2500 BCE, smiths from Sumeria (modern-day Iraq)

Early hunting weapons with cutting blades were often made from sharpened stone, such as this Native American arrowhead.

Roman soldiers used the testudo (tortoise) formation during sieges. The platoon is protected from attack from above and in front by the leather-covered shields. The shield was an important piece of defensive equipment until the invention of gunpowder.

were able to craft fairly sophisticated helmets from bronze, and their effectiveness forced changes in the design of the mace. In response, the head of the mace became more oval in shape, so that the force of a blow was concentrated on a specific point, causing greater damage to the helmet. Gradually the mace evolved into the battleax. Unlike the mace, the ax was not merely reliant on its weight to inflict damage but had a cutting edge as well.

ARMOR

As early as 1500 BCE, the Egyptians were making body armor by covering leather clothing with small pieces of bronze. The Assyrians were the first to use iron to make such armor. Greek artifacts from the 3rd century BCE provide the earliest evidence of mail, flexible body armor made from small overlapping or linked metal pieces. There is some evidence that mail was used even earlier in Celtic Britain. By the 1100s CE suits of mail armor covered the entire body. Improvements in the metalworking skills of European armorers in the 13th century resulted in the production of cheaper and stronger iron. As a result, chain mail was gradually replaced with plate armor, which was made from larger pieces of metal. At first plate and mail were used together. However, the gaps between the plates created vulnerable spots and led to the creation of complete suits of armor in Germany and Italy in the 14th century.

A history enthusiast tests out fighting in plated armor.

The ax remained the most important cutting-edge hand weapon until around 1200 BCE. Around this date, people started to cast swords from iron. While swords had been made before this date, they were cast from bronze, which could not be made into an effective strong, long, and sharp blade. It was only after the development of iron smelting that the sword came into its own. While its design changed repeatedly over the next 3,000 years, it was used successfully in battle right up until World War I (1914–1918).

Slings and Arrows

The sword, ax, and mace are all shock weapons, which are used in hand-to-hand combat. Equally important were missile weapons, which were used to attack the enemy from afar. The most basic of these were those thrown by hand, such as stones, spears or darts, which relied on human muscle power alone to generate speed. A further development was the sling, which consisted of two thongs attached to a pouch. A small stone or lead shot was put in the pouch. The thrower whirled the sling around his head before letting go of one of the thongs, and using the speed generated by the whirling

Although a sling is easy to make, a stone from a sling was seldom deadly unless it was in the hands of an expert.

FACTS AND FIGURES

- The longbow was 6 ft (180 cm) long and could propel a 3 ft (90 cm) arrow up to 1,200 ft (365 m). A skilled archer could fire up to 10 arrows in 60 seconds. Some special arrows with hardened steel heads could penetrate plate armor and iron mail from 300 ft (90 m) away.
- The crossbow was 3 ft (90 cm) long and could propel a 1.5 ft (45 cm) arrow 900 ft (270 m). It was accurate and simple to use, but a skilled archer could fire only about two arrows a minute.

Weapons of medieval warfare. The infantry are using crossbows (left) and longbows; the cavalry have swords and lances.

Missile weapons used in medieval warfare: the throwing spear (left), a short bow (center), and the more powerful crossbow.

THE COMPOSITE BOW

The effectiveness of the archer in battle was increased by the invention of the composite bow around 3000 BCE. This weapon had a long range and considerable penetration power.

Composite bows were made by binding animal sinews (fibrous body tissues) to the front and horn to the rear of a curved wooden stave. The outer arms of the bow bent away from the archer, forming a reverse curve. The power of the bow relied on stringing the bow so that the layers of horn and sinew were bent back toward the archer (stretching the sinews and compressing the horn); this built up extra tension, which propelled the arrow forward with an enormous amount of force for such a small weapon.

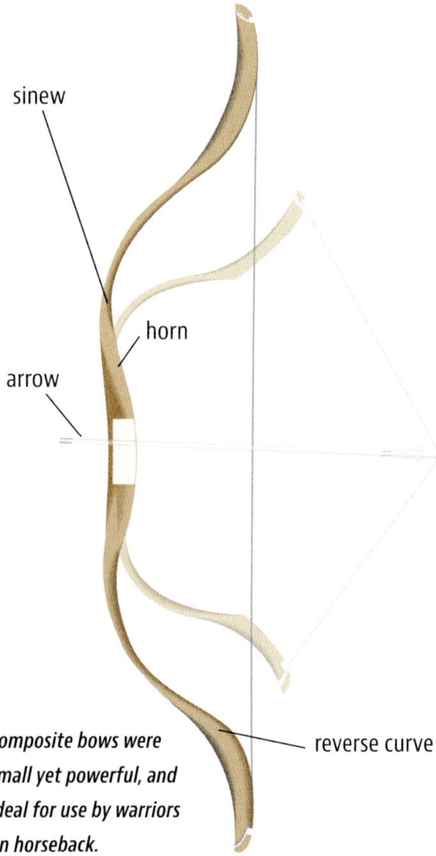

Composite bows were small yet powerful, and ideal for use by warriors on horseback.

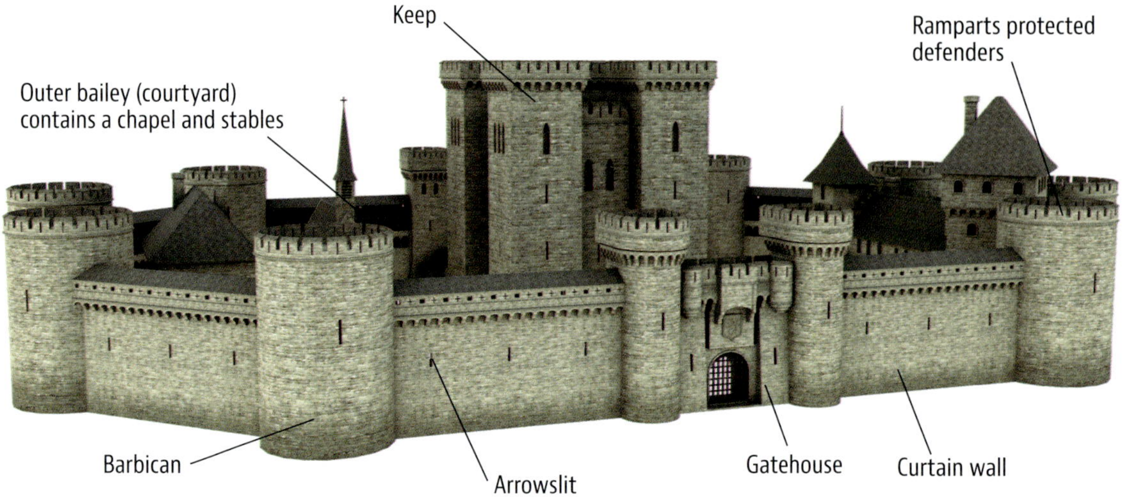

A European stone castle from the Middle Ages

motion to launch the stone toward the enemy. In around 300 BCE, slingers played an important role in armed conflict.

The most important missile weapon was the bow, which was first used around 30,000 BCE. Bows use the tension that is stored in the limbs of the weapon when they are pulled back by a bowstring to propel arrows forward at considerable speed. One of the most important technical developments was the introduction of the composite bow around 3000 BCE. The composite bow (made from a mixture of wood, animal sinew, and horn) continued to be a highly effective weapon throughout the Middle Ages. It was used to devastating effect by the forces of the Mongol emperor Genghis Khan.

In the 12th century two further developments increased the power of the bow as a weapon. The first was the rediscovery of the crossbow. The idea of placing a bow laterally across a wooden stock and fitting it with a mechanical winding mechanism and trigger was first explored by the Chinese in the sixth century BCE, but the design did not spread to Europe until 1,800 years later. The crossbow's winding system

ATTACKING A CASTLE

When the Greek ruler Alexander the Great invaded the Persian Empire in 334 BCE, his engineer Diades invented a machine called the crow. This was a long, heavy bar suspended from a vertical frame that was used to knock down the upper sections of a wall. Diades also invented another siege weapon called the telenon. It was a large basket slung from a boom on a vertical frame that could contain soldiers. The basket was raised and then lowered to place soldiers directly onto the enemy's walls.

Soldiers attacking a castle of the enemy used a telenon to lift them up to the castle ramparts.

TYPES OF SIEGE ENGINE

Sambuca
The sambuca (siege ladder) was designed by the ancient Greek engineer Damis of Colophon. Soldiers climbed a ladder to enter the top compartment. Stones and rocks were then loaded into boxes at the rear, raising the men at the front up to the level of the battlements.

Fire-raiser
Fire-raisers were long wooden beams that carried cauldrons of lighted coals, sulfur, and pitch. These were used to start fires in the enemy's stockade. The wooden beam was hollow with a central iron tube and one iron-clad end to prevent it catching fire. A soldier pumped bellows at the other end to keep the fire alight.

Battering Ram
The machine illustrated here was used in the 4th century BCE. The ram is metal plated and housed in a protective shell made of fire-resistant compressed seaweed and ox-hide. The machine was pushed forward on its wheels to the base of the wall and the ram was swung backward and forward using ropes and pulleys.

SAMBUCA

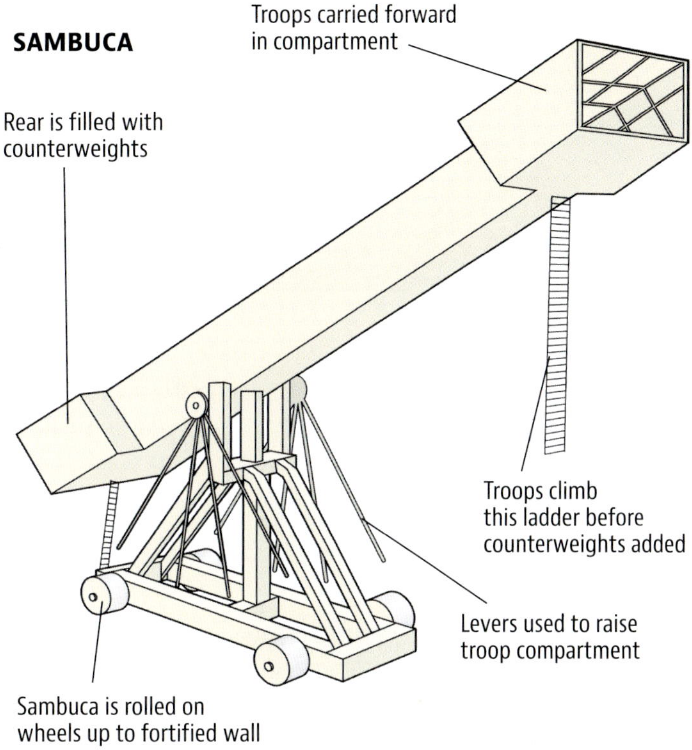

BATTERING RAM

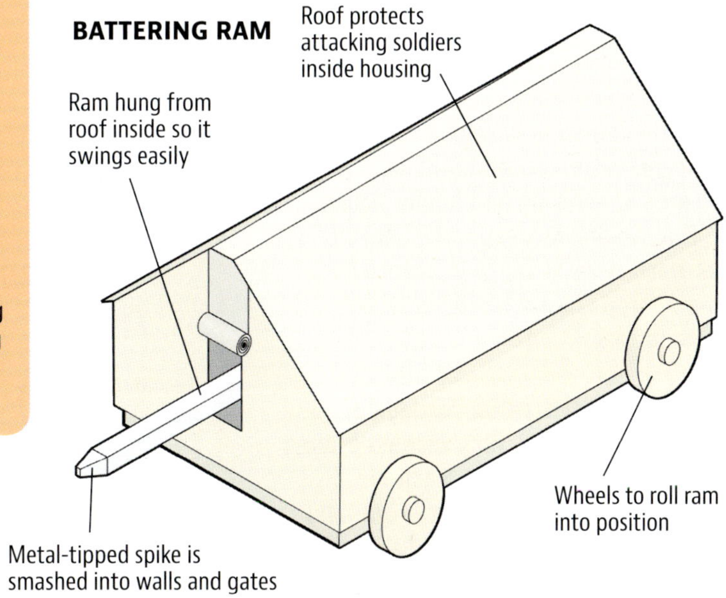

This papyrus artwork shows an Egyptian war chariot. The archer on board can move around the battlefield at speed.

gave it enormous power, and the iron bolts it fired were able to pierce even the strongest armor. However, the crossbow did have its drawbacks, the most notable of which was the time it took to reload. After each firing, the string had to be gradually winched back to create the necessary tension to power the next projectile.

These problems were overcome by the Welsh longbow. The longbow was made from yew, a wood that is extremely strong but also flexible and so it had the extra stretchiness of a composite bow. At around 6 ft (180 cm) in length, the longbow was extraordinarily powerful and extremely accurate, but only in the hands of a strong and highly skilled archer. The use of the weapon by English soldiers was a pivotal factor in victory over French mounted knights at the battle of Agincourt in 1415.

Fortifications and Sieges

People began to build defensive walls around their homes and settlements tens of thousands of years ago—fortifications have been found at Jericho in the Middle East that date back to around 8000 BCE. Until 900 BCE attacking armies lacked a reliable way of assaulting stone forts and surrounded them instead, forcing the occupants to come out fighting or surrender from starvation. Around that time, however, the Assyrians (from present-day northern Iraq) developed techniques to attack forts. They relied on machines such as the battering ram. Built on a wheeled wooden frame and covered by protective plates of specially treated leather, the Assyrian battering ram was swung on ropes and its impetus smashed through defensive walls.

The Assyrians also went over the walls using scaling ladders and ramps. Mantelets (large shields sometimes mounted on wheels) protected advancing soldiers from archers and slingers within the fort.

BOOSTING FIREPOWER

Improvements over the centuries boosted the firepower of the ancient world's mechanical artillery. The degree to which the arms of artillery machines could be bent backward increased. This, in turn, increased the amount of torsion in the rope springs to which they were attached, thereby increasing the machine's firepower.

1 The arms of an early Greek arrow-firer could be moved 35° from a starting point of 20°.

2 The curved arms of the scorpio from c.50 BCE allowed them to be moved an extra 10°.

3 Around 100 CE the cheiroballistra had a metal frame, which allowed the angle of twist to be increased since the springs could be placed farther apart.

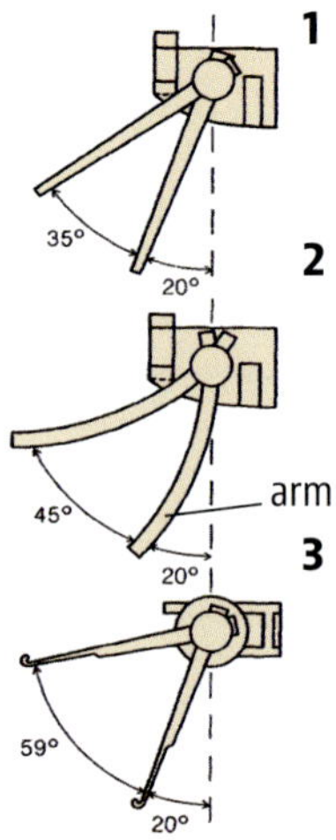

ONAGER CATAPULT

The Roman catapult was known as an onager, or "wild ass," because of the heavy recoil kick. The weapon's power came from the torsion of twisted ropes that acted as a giant spring. The arm was cranked back by the lever.

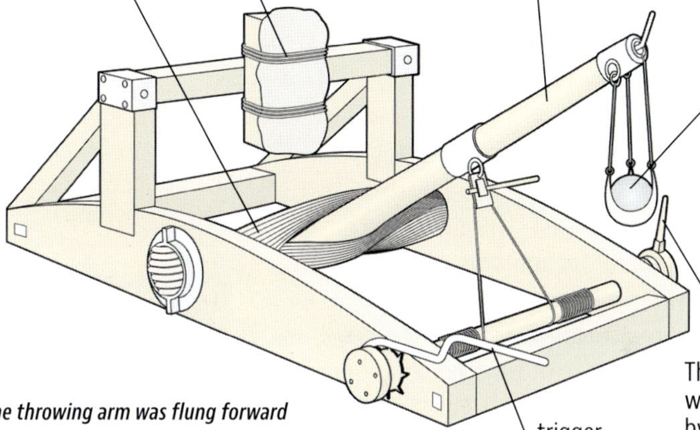

The onager fired stone shot. The throwing arm was flung forward by a coiled spring of ropes. The weapon dates from around 350 CE.

Mechanical Artillery

Ancient artillery consisted of machines used for hurling heavy missiles. The invention of mechanical artillery is traditionally attributed to the tyrant King Dionysius the Elder (405–367 BCE) of Syracuse, on the coast of the island of Sicily. However, large stone-throwing catapults and machines (ballistas) capable of firing giant arrows a distance of 1,500 ft (450 m) were first used as early as 500 to 400 BCE. Dionysius made effective use of such machinery in his conquest of southern Italy between 379 and 367 BCE. The Macedonians produced lighter versions of the catapult and ballista that could be used on the battlefield.

Cavalry and Chariots

The wheeled chariot first appeared in Mesopotamia (present-day Iraq) in 3000 BCE. By 1800 BCE, the light two-horse chariot was in use. It had spoked, hubbed wheels that turned around a fixed axle and carried two people. Ancient armies used chariots combined with infantry (foot soldiers), archers,

BALLISTA

The Roman ballista shown here dates from around 25 BCE. It looked like a large crossbow but was a torsion weapon, using two sets of tightly twisted ropes to provide the propelling force. To load the ballista, a sliding trigger was pushed forward until it could hook onto the firing string. Then the trigger was wound back by a lever, pulling the string very tight. A ratchet prevented the string from flying forward until released.

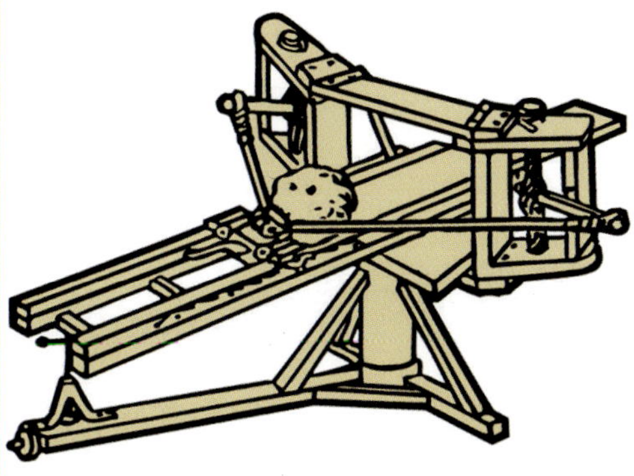

and spearmen. With an archer on board, the chariot was a mobile platform of firepower. Chariots were most important in warfare between 2500 and 900 BCE.

Chariots were more effective than early cavalry because without stirrups cavalrymen could not swing heavy weapons nor easily resist being pulled from their mounts. Between 400 and 1300 CE, however, cavalry evolved into the dominant arm of battle in both Europe and the Middle East. This dominance was based on several technological factors.

Most crucial was the invention of the stirrup and related technologies like the solid-tree saddle and bridle with bit. The solid-tree saddle spread the weight of a rider across the whole back of the horse while the stirrup effectively welded the rider to the horse. This made it easier for a mounted soldier to charge into the enemy, while the bridle with bit enabled him to maintain control over the horse while fighting. Stirrups probably originated in the Central Asian steppes between 300 BCE and 100 CE. They were first used in Western Europe between 500 and 1000.

Improvements in metalworking helped reinforce the dominance of cavalry and led to the development of suits of heavy armor, worn by upper class warriors known as knights. Swords became longer, up to 40 inches (1 m) in length, while by 1050 the spear was replaced by the lance, a long wooden shaft with a sharp metal spearhead, which was used for knocking enemy cavalry off their horses.

Horse-borne troops continued to dominate warfare in Western Europe until the mid-14th century, when the emergence of the crossbow and the longbow tilted the balance of power in favor of the foot soldier. Another important factor in the cavalry's downfall was the reemergence of the pike in the 15th century. The pike was a wooden staff, typically about 12 ft (3.5 m) long, fitted with a long, sharp point. The weapon was held in battle, not thrown, and could unseat charging riders. Pikes and other hand-held weapons such as halberds and long axes were used by the

ELEPHANTS IN WAR

The use of elephants in warfare started in India. It spread to Greece and Carthage (now Tunisia, in North Africa). Elephants inspired fear in troops and horses, but enemies learned how to make them stampede, causing danger to the armies using them. Anti-elephant measures included iron spikes chained to the ground to pierce the animals' feet. As cannons and guns became widespread, the use of elephants declined.

This stone carving from Cambodia shows elephants marching into battle alongside foot soldiers.

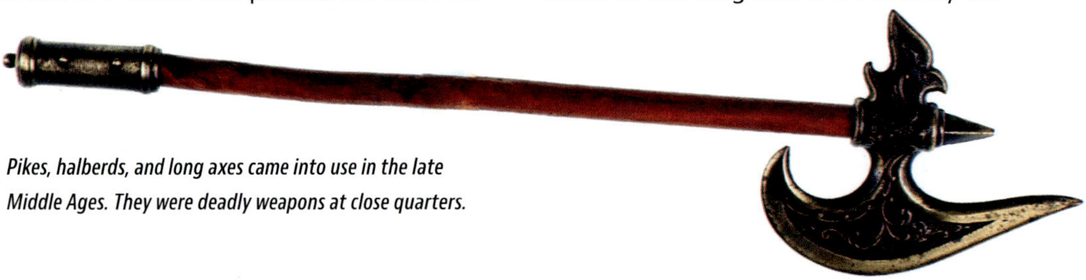

Pikes, halberds, and long axes came into use in the late Middle Ages. They were deadly weapons at close quarters.

Swiss infantry to inflict a number of heavy defeats on the knights of Burgundy and the Holy Roman Empire between 1460 and 1499.

Gunpowder

Tthe invention of gunpowder completed the end of the success of knights in warfare. Gunpowder's origins are not really known. A formula for fireworks was provided in an 11th-century Chinese manuscript. The formula was for a slow-burning substance, however, and was not the explosive of today. The Chinese used it for filling rockets. In China these rockets were used to drive away evil demons but were probably also used in warfare in the 12th century. Soon, gunpowder was used in early grenades.

The German alchemist and monk Berthold Schwarz is said to have discovered the explosive by accident somewhere between the years 1313 and 1353. However, it is now clear that explosives came from China, where they had been invented centuries earlier.

SOCIETY AND INVENTIONS

The Importance of Knights

The military importance of the armored cavalry, often known as knights, was such that monarchs were prepared to give land to knights in return for military service. Training for the skills needed to fight as a knight usually started in childhood and continued throughout life. There was a strict social hierarchy: agricultural land was worked by peasants, and the knight lived off the surplus. This allowed a knight to purchase expensive equipment and keep a string of horses and his military assistants, while showing off his skills at tournaments (shown in a modern reconstruction, *right*). Eventually, access to this warrior class became based solely on heredity (birth). As a result, knights became part of an aristocracy that controlled Europe for around 1,000 years.

Early grenades were bamboo tubes filled with powder as well as stones, broken porcelain, and iron balls. We know that Chinese troops were using bamboo guns in the 13th century, and that by the 14th century they were making metal guns.

Knowledge of China's black powder, as it was known, reached Europe, probably via the Mongols, in the 1200s. Nearly another 100 years passed before gunpowder was used as an explosive propellant in European warfare. Two Europeans, English scientist Roger Bacon (1214–1294) and English scholar Albertus Magnus (c.1200–1280), described in detail the making of gunpowder. The first recorded Western use of gunpowder was at the battle of Crécy, in northern France, in 1346.

The first European gunpowder weapons were small metal pots packed with powder that was then ignited to fire arrows. By the late 14th century enormous versions of these metal pots, called bombards, fired stones weighing as much as 300 lbs (140 kg).

Developments in metalworking allowed the production of cannons with longer barrels, which had more range and power. Around 1450 cast-iron balls were invented. Because of their size and weight cannons were first used primarily in siege warfare. By the 1490s, the French had developed light bronze-cast

LIGHTER CANNONS

King Gustavus II (1594–1632) of Sweden made great progress in improving cannons. Before his time, cannons were too heavy to move once they were on the battlefield. Gustavus reduced the weight of the guns by making barrels shorter and thinner, but maintained their strength by casting them from a lighter mixture of copper and iron. The smaller cannons could be moved around the battlefield to assist infantry and cavalry attacks.

SHRAPNEL

In 1784, a British army officer named Henry Shrapnel (1761–1842) invented the shrapnel shell. This was a canister filled with explosives and bullets. When it was fired from a cannon, it exploded in the air, spraying metal over a wide area. This was a deadly weapon against both infantry and cavalry.

Two gunners aim a cannon using clinometers, devices for measuring the angles of inclines, to aim the gun accurately.

cannons. These were mounted on two-wheeled carriages pulled by horses.

Handguns

The earliest handguns were developed at the same time as cannons. These were simple tubes of iron or brass with a hole in the top. Powder and ball were placed in the tube, the firer then held the gun with one hand while igniting the powder through the touchhole on top with a red-hot iron. As well as being very noisy, these small hand cannons were also

MATCHLOCK MUSKET

One of the first handguns was the matchlock musket (*below*), invented by the Spanish military in the 1530s. Although it was difficult to handle and had to be rested on a metal fork when fired, its long barrel made it powerful enough to penetrate all types of armor. The matchlock also took a considerable time to reload once fired: a soldier had to perform 97 separate movements to fire the gun again. This resulted in a low rate of fire, perhaps just one shot every two minutes. In spite of these drawbacks, the matchlock musket became the main firearm of European armies in the 17th century. King Gustavus II of Sweden simplified the loading process with a paper cartridge that contained gunpowder and a musket ball. This improvement doubled the rate of fire.

THE BAYONET

The advent of firearms presented European armies with a dilemma. Once fired, the time taken to reload made the gunner vulnerable to attack by other soldiers armed with traditional weapons such as the sword, ax, or pike. Infantry units needed soldiers using different types of weapon.

Warfare was eventually changed by the combination of two technologies. In France (in the town of Bayonne) someone invented a long knife that fitted into the muzzle of a hand gun. This "plug" bayonet gave the gunner a deadly thrusting weapon, but meant that he could no longer fire.

In response to this it is believed that French military engineer Marquis de Vauban (1633–1707) constructed the socket bayonet. This was attached to the outside of the barrel and allowed the gun both to be fired or used as a minipike for close-quarters fighting. Now, all soldiers could fight both at long range and also hand-to-hand.

The musket-bayonet combined two technologies: gunpowder for long-range firing and a steel bayonet for close fighting.

inaccurate, and their barrels would become too hot to hold after only a few shots had been fired.

In the 15th century improvements were made in the ignition and accuracy of these firearms, and they became a class of weapon distinct from cannons. The touchhole was moved from the top to the side, and a small pan was added to hold priming powder. A cord was soaked in saltpeter (potassium nitrate, a flammable powder extracted from rocks) so that it continued to smolder after being lit. This was used to ignite the weapon. This meant that the gunner did not have to stand close to a fire in order to use his weapon.

Ignition and firing systems were further improved with the development of the matchlock. This was an S-shaped mechanism attached to the side of the gun that held the burning cord and, when depressed by means of a trigger, ignited the powder. The barrel was placed within a wooden stock, and a curved wooden butt was added to the end of the barrel. This meant that the soldier firing the gun could hold the wooden butt piece against his shoulder while aiming and firing the weapon in the same way that a soldier would fire a rifle today. This gun was called a harquebus, meaning hooked tube. The harquebus was the standard infantry weapon until the middle of the 16th century.

At first pikemen defended hand gunners while they reloaded, but by the end of the 17th century, the bayonet, a dagger that attached to the muzzle of a firearm, meant gunners could defend themselves. Meanwhile,

PERCUSSION CAP

In 1815, the American inventor Joshua Shaw (1776–1860) was credited with devising the percussion cap by enclosing the priming powders in a metal tube. The cap was detonated by a hammer on the musket striking a pin within the tube. The percussion-cap musket was much more reliable than the flintlock musket and even fired during rain. Despite this step forward in handguns, problems still remained. For example, it still took a soldier a long time to ram a ball down the barrel of his gun. For this problem to be overcome, the handgunner had to await the arrival of a new technology.

MATCHLOCK AND FLINTLOCK

Matchlock

The matchlock firing mechanism was developed throughout the 15th and 16th centuries. It used an S-shaped mechanism called a serpentine, which held a slow-burning saltpeter cord, or match. The serpentine was pivoted. Pushing the trigger lifted the bottom of the mechanism and brought the match down, bringing it into contact with the priming powder that then ignited the powder in the barrel.

MATCHLOCK

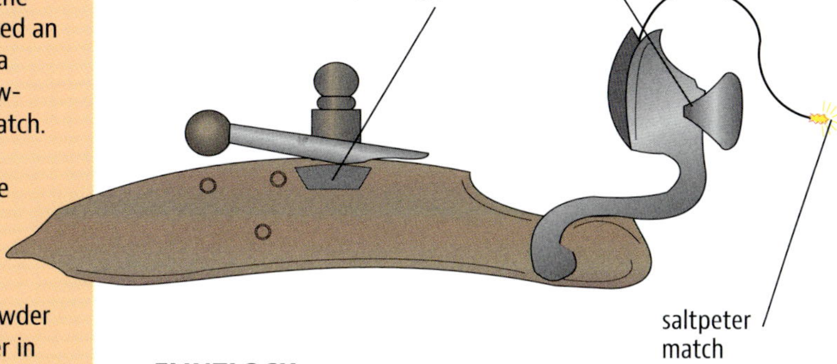

Flintlock

The matchlock was replaced by the flintlock in the 17th century. The flint was held in a vice known as a cock. When the trigger was pulled, the cock was propelled onto a piece of steel called the frizzen. The blow not only produced sparks as flint hit steel, it also knocked the frizzen forward, uncovering the powder in the pan and allowing this powder to be lit, which ignited the powder in the barrel.

FLINTLOCK

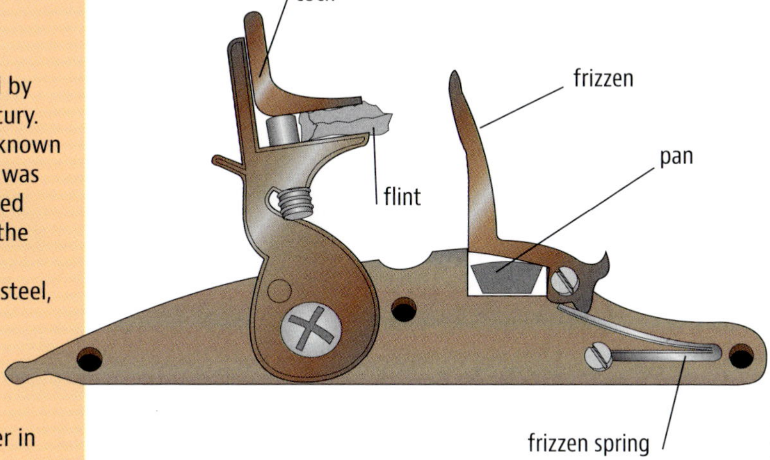

the firepower of the infantryman was increased as the matchlock musket was replaced by the flintlock musket at the end of the 17th century. Flintlock muskets were quicker to load, cheaper than other mechanisms, and easier to repair since they had fewer small parts.

The development of gunpowder weapons meant that a key part of soldiering was now using technology rather than developing hand-to-hand fighting skills. Knowing about the best angles for firing cannon at fortifications or how to make and transport gunpowder became crucial skills. People who had the skills to control the new "engines of war" became known as engineers, a word that became widely used in military circles. Not until the 18th century did British inventor John Smeaton (1724–1792) call himself a "civil"engineer.

1850 TO THE NUCLEAR AGE

The huge mortars used during the Siege of Yorktown in 1862 weighed 10 tons (9 metric tons) each.

At first, the Industrial Revolution (c1760–1840) in the United States and Europe did not have much effect on the way armies were armed or fought battles. But by the American Civil War (1861–1865), it was clear that industry was changing the nature of warfare forever.

The first weapon to be changed dramatically was the infantry (foot-soldier) rifle. In Prussia (now part of Germany) Johann Dreyse (1787–1867) invented the first modern breech-loading rifle, the "needle gun," which entered military service in 1848. However, the design of the weapon was a closely guarded secret, and at the start of the American Civil War in 1861 most soldiers were still using the older muzzle-loading rifles. It was during this war that the breech-loading rifle became widespread. Tyler Henry, an American gunsmith, introduced the Henry repeating rifle, while Christopher Spencer invented a short-barreled rifle, or carbine. In both of these breech-loaders, ammunition was stored in an attached magazine, which allowed rapid firing. A modified version of the Henry rifle, called the Winchester, became common in the 1870s.

LOADING A RIFLE

Muzzle-loading

Up to the 1860s, most infantry carried weapons that were muzzle loaders. Powder was poured down the barrel, and then a lead ball or bullet was rammed down on top of the powder and kept in place by a plug of cloth. It was difficult for a soldier to fire more than three shots per minute using this method.

Breech-loading or Bolt-action

The development of precision steel parts opened the way for breech loading. In a breech-loading rifle, the powder and ball are inserted as a combined cartridge into the breech of the rifle, which can be opened by sliding back a steel cylinder, called the bolt. The bolt is then pushed forward to seal the breech, the gun is fired, and another cartridge can be inserted.

Self-loading Rifle

The next step forward was the self-loading rifle. This works by using some of the exhaust gases from firing to drive a reloading mechanism. The bolt is driven back either by the gas itself or by a gas-operated piston. The spent cartridge is ejected and a new one loaded from a magazine as the spring-loaded bolt returns to a forward position.

Assault Rifle

Modern assault rifles are capable of automatic fire: the trigger does not need to be released between shots. They use the exhaust gases in the same way as a self-loading rifle but the firing is continuous. The main problem with automatic rifles is in slowing the rate of fire to a controllable level.

A Kalashnikov AK-47, one of the most widely used modern assault rifles, was first developed during the late 1940s.

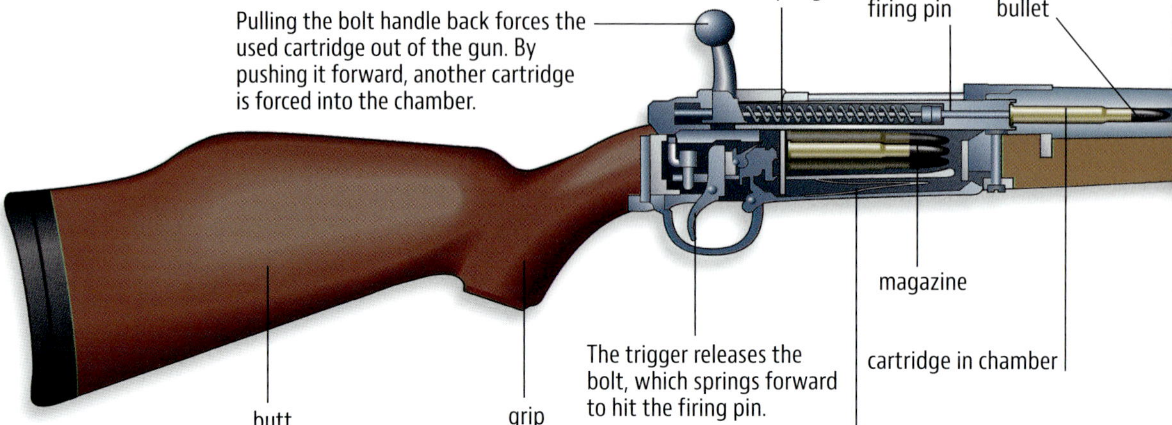

BOLT-ACTION RIFLE

A World War I rifle could hold 10 rounds, or cartridges, in its magazines and could hit targets up to 2,000 ft (600 m) away.

In the 1880s, workers at the Springfield Armory in Massachusetts mass-produced revolvers that could be loaded with six bullets.

By the 1880s, the Springfield Armory in the United States, Enfield Arsenal in Britain, and Mauser factory in Germany had all produced fairly similar robust infantry rifles. They were magazine-fed, breech-loaders, generally of about 0.3 inch (7.62- mm) caliber, which referred to the diameter of the inside of the gun barrel. They could fire 10 aimed shots per 60 seconds. These weapons were standard infantry rifles in both world wars.

GUN METALS

The technique of blasting molten pig iron (raw iron) with air to produce high-quality steel was invented by British metallurgist Henry Bessemer (1813–1898). It changed the way guns were manufactured. Using steel, guns could be made stronger, lighter, and more powerful. Manufacturers like Alfred Krupp (1812–1887) in Germany used tougher steels to produce breech-loading, rifled artillery pieces that were far superior to the older designs.

THE BENEFITS OF RIFLING

Most firearms made before the 1840s were of the smoothbore type—the inside of the barrel was simply a smooth cylinder. These weapons fired loose-fitting bullets that could be pushed easily from muzzle to breech. This meant that the bullet tended to wobble as it traveled up the barrel after the gun was fired, reducing the gun's accuracy.

The advantages of rifled weapons, which have a series of coiling grooves inside their barrels, was already known. The grooves make the bullet spin as it leaves the barrel, leading to a more stable flight and greater accuracy.

For this effect to work, however, the bullet must fit tightly inside the barrel, making it difficult to ram from muzzle to breech and reducing the rate of fire.

A solution to this problem was found in 1841 by Claude-Etienne Minié (1804–1879). He developed a cone-shaped bullet that was smaller than the diameter of the barrel. This allowed the bullet to be dropped down the barrel. When fired, the gases produced in the barrel caused the bullet to expand, fitting it into the grooves of the rifle. This ensured great accuracy. Rifled weapons are now standard in infantry weapons and artillery.

The rifled barrel of this train-transported WWI artillery gun made shells spin so that they were more likely to hit a distant target.

ALFRED NOBEL

Alfred Nobel (1833–1896, *below*) was a brilliant Swedish engineer and chemist. He built up a large manufacturing business. Nobel was, fascinated by the properties of explosives and began intensive and dangerous research to make more effective explosives than the commonly-used gunpowder.

Experimentation with guncotton and nitroglycerine led him to discover several new explosives that were stable, safe, and very powerful. Among these were dynamite and blasting gelatin, which are used in industry, and ballastite, which became the basis for most modern ammunition.

Nobel was treated with suspicion by many Swedish people for his research into deadly explosives. He left much of his fortune to a foundation that established five open competitions in areas of human endeavor. One of the most famous of these prizes is the Nobel Peace Prize.

Alfred Nobel in his laboratory

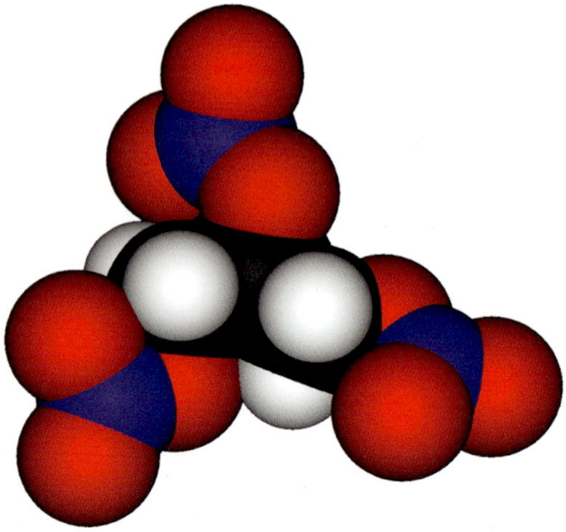

A molecule of nitroglycerin has nitrogen atoms (blue), oxygen atoms (red), carbon atoms (black), and hydrogen atoms (white). Its chemical formula is $C_3H_5(NO_3)_3$.

In the 1930s attempts were made to further improve the firing rate of infantry rifles. The Canadian-American engineer John Garand (1888–1974) developed a self-loading rifle. His design was adopted by the U.S. Army in 1936, giving American soldiers the most modern infantry weapon available at the time. During World War II (1939–1945) the Germans developed the MP-44, the first assault rifle—accurate and capable of single-shot and automatic fire. This design was copied by the Russian engineer Mikhail Kalashnikov (1919–2013) to produce the AK-47 (see the box on page 19).

Improving Artillery

Artillery pieces did not change as quickly as the rifle during the 19th century, but as infantry muskets began to be rifled, so gun manufacturers tried to rifle artillery pieces. The problems associated with rifling heavy weapons were considerable, and before much progress could be made a method of manufacturing stronger gun barrels had to be found. In 1854, William Armstrong (1810–1900),

Much of the work at the Nobel dynamite factory was done by women, who mixed dangerously volatile nitroglycerin with earth and sawdust to make it more stable.

a British engineer, made the first reliable rifled artillery barrels. Constructed from a series of forged iron tubes, Armstrong's barrels were quickly adopted by the British Army.

The real breakthrough for artillery came in 1897, when the French Schneider company produced a new and radical 3 inch (75- mm) gun. This weapon had a rifled barrel and was breech loading. It also had a liquid-filled recoil system that returned the barrel to its original position after firing. Because it did not have to be moved between shots, the "75" could fire six aimed shells per second—far faster than any other artillery piece at that time. Although artillery pieces would be developed further during the 20th century, with higher rates of fire, longer ranges, and heavier explosive charges, they all followed the pattern of the French 75.

Hand-in-hand with the inventions of new rifles and guns came the development of new, more powerful explosives. Gunpowder had been used for centuries, but its explosive force was relatively limited. In 1846, Austrian chemist Christian Schonbein (1799–1868) produced guncotton by adding nitric acid to cotton. Guncotton was far more powerful than gunpowder, but it was also highly unstable and dangerous and was not adopted

POWERFUL EXPLOSIVES

While ballastite and cordite became the main explosives used for firing guns, the explosive most widely used to fill shells was picric acid, first tested by the French in 1886. A safer, more stable alternative to picric acid, trinitrotoluene (TNT), was discovered in Germany in 1902 and became the standard explosive for shells. TNT is much more powerful than gunpowder. Long-range artillery bombardments of terrible destructiveness became a feature of both world wars. During WWII, the British company ICI developed the first plastic explosives. These were safe to handle, easy to mold into any shape, and difficult to detect. First used by British commandos and resistance groups in WWII, plastic explosives have since been used by terrorist groups around the world.

FLAME WEAPONS

Flamethrowers were developed by the German Army in the early 1900s and were used against Allied troops in 1915, during World War I. The original flamethrower was a steel tube that squirted burning oil using compressed air. The British soon developed a similar device, but all flamethrowers in Word War I were clumsy and short ranged.

Napalm (jellied gasoline) was developed by U.S. scientists in 1942. It was first used in 1944, during World War II. Napalm burned with intense heat and clung to whatever it touched. The British Crocodile tank, which towed an armored trailer filled with napalm, could fire 100 shots to a range of 260 ft (80 m).

During the Vietnam War (1955–1975), the US military dropped napalm bombs to clear large areas of land.

A bomb test shows the destructive power of napalm bombs. The flames burn at 1,500°F (800° C) and cling to whatever they touch, including flesh.

by many armies. However, the Swedish chemist Alfred Nobel made a breakthrough in 1875 by combining guncotton with nitroglycerine to make dynamite, a substance that is stable until detonated and burns without smoke. From dynamite, which was used mainly in mining and industry, Nobel developed ballastite, a military explosive. English chemists Frederic Abel and James Drewer were working along similar lines. They produced cordite, another stable, smokeless explosive.

Machine Guns

Modern artillery was one of the dominant weapons in the world wars. The machine gun was another. The idea of a repeating gun had been around for a long time, but it was only in the 1860s that practical designs were used in battle. In 1862, the American inventor Richard Gatling (1818–1903) built the first reliable machine gun. It was hand-driven and consisted of 10 barrels rotating around a central crank. During a full rotation, each barrel was loaded and fired and the spent cartridge was ejected. The French also developed a hand-driven machine gun, the Mitrailleuse, which fired 125 rounds per minute. This was used, but to no great effect, in the Franco-German War of 1870.

With the invention of Nobel's more powerful smokeless explosives, the American inventor Hiram Maxim (1840–1916) went on to develop the first fully automatic machine gun

Hiram Maxim demonstrates his machine gun in the 1880s. Maxim also invented the gun silencer and is credited with the development of an improved mousetrap.

in 1884. Instead of being hand-cranked, the Maxim gun used the gas and recoil produced by firing to load the next round. Maxim's gun could fire 500 rounds per minute. The Maxim gun was copied by most European armies and

HOW A MACHINE GUN WORKS

A general purpose machine gun can perform a wide variety of roles, such as providing supporting fire or shooting down aircraft, but is light enough to be transported by infantry. The MG34 developed by the German Army in 1934 was the first weapon of this type, and its design has been widely copied ever since. These guns can be mounted either on a bipod (two-legged) stand or on a tripod (three-legged) stand for shooting up into the sky. Bullets are fed into the breech via flexible belts. A gas-operated reloading piston ejects the used cartridge and loads the next bullet automatically. Such guns can fire over 1,000 rounds per minute but always fire in short bursts to prevent the barrel from overheating, and the barrels have to be changed often. A flash hider at the end of the barrel makes it more difficult for the enemy to locate the gun.

M249 MACHINE GUN (1984–PRESENT DAY)

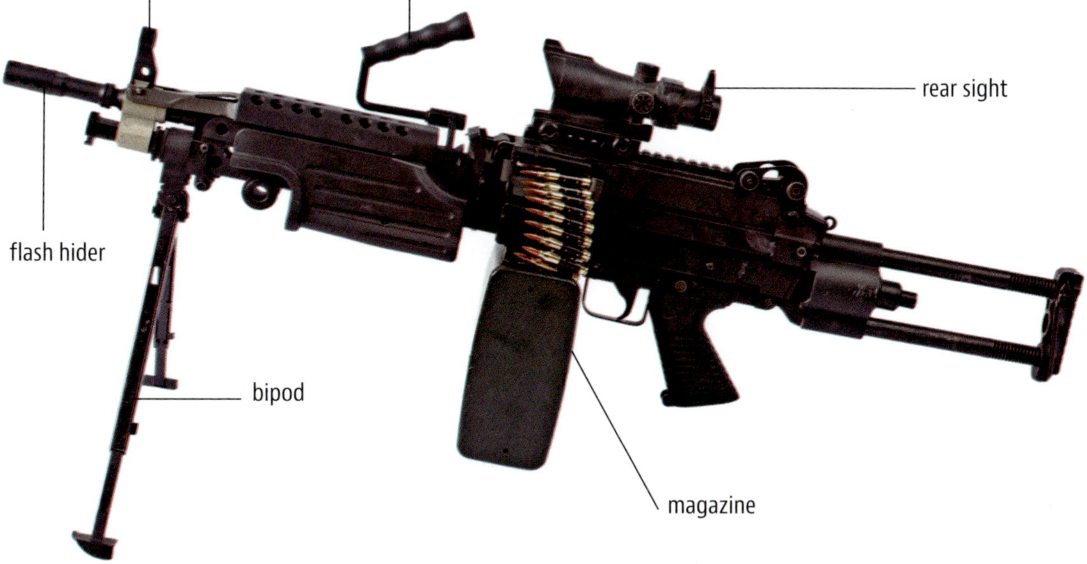

was used to deadly effect in World War I (1914–1918). To prevent the barrel from overheating during sustained firing, the Maxim gun was cooled with water, which made it heavy and difficult to transport on the battlefield. This led other inventors, notably Isaac Newton Lewis (1858–1931), to develop lighter air-cooled machine guns. The Lewis gun, invented in 1912, was adopted by the Allied armies in World War I and gave soldiers a machine gun they could carry.

Grenades and Mortars

Another simple infantry weapon that was widely used during World War I was the grenade, a hand-held explosive that could be hurled at the enemy, detonating after five or seven seconds. Grenades gave soldiers "pocket artillery," which was of great use in trench warfare.

Mortars were also developed to help soldiers in the trenches. Mortars (short-

This early 18th-century mortar, known as a "partridge," fired one large shell plus a dozen smaller grenades from side barrels.

SOCIETY AND INVENTIONS

The Cost of War

All of the developments in weaponry during the 19th and 20th centuries resulted in armies possessing vastly increased killing power. Riflemen could fire at least 10 rounds a minute with great accuracy. The machine gun, described as the "concentrated essence of infantry," could deliver high volumes of fire—advancing infantry were literally mown down. Artillery pieces firing explosives and gas killed thousands of people. On July 1, 1916, the British Army suffered 60,000 casualties in one day. Over the course of WWII, 32 million people died, as warfare became more mechanized and weapons more powerful. In the 20th century, wars regularly killed several percent of a country's entire population.

An American WWII military cemetery in Luxembourg. The weapons of modern warfare create many casualties among soldiers and also civilians.

MINES IN WARFARE

The first mines were large bombs placed under an enemy position by miners, who dug tunnels under the battlefield. In response to the threat of tanks, inventors in the 1930s developed modern mines. These are buried containers filled with explosives and detonated by the pressure of a tank or person above them. Mines are easy to make and lay but difficult to detect and remove. Clearing mines from areas affected by warfare is still a deadly problem for many countries worldwide.

A sign warning that land mines have been laid in the area.

barreled, high-angled guns for dropping shells onto the enemy) had been in use for many years, but the first modern mortar, called the Minenwerfer, was designed by the Germans in 1908. The British and French armies produced their own versions, the Stokes and Brandt mortars. The German 3.2 inch (81 mm) mortar of 1939 was an extremely effective design consisting of a simple tube mounted on adjustable legs and resting on a base plate. The operator dropped the bomb down the tube to fire the mortar. All modern mortars follow this basic design.

Chemical Weapons

One of the most terrible weapons developed during WWI was poison gas. The Hague Convention of 1899 condemned the use of gases in warfare, but the Germans, French, and British all continued research. The German chemist Fritz Haber (1868–1934) developed poisonous chlorine and phosgene gases for military use. The French had used tear gas against the Germans in August 1914, but it was only when poisonous chlorine gas was used

Grenades are primed by pulling out a pin. The thrower releases the lever, which sets off a fuse and the grenade explodes seconds later.

British children wear gas masks during World War II as a defense against chemical-weapon or gas air raids. No gas attacks were made on the United Kingdom.

by the Germans near Ypres in Belgium on April 22, 1915, that gas had a major impact on the war. Without gas masks, many of the French soldiers panicked, leaving a gap in their defensive line. Initially, poisonous gases were released from cylinders and spread by the wind, but soon shells were filled with gas and fired at enemy positions. During WWI gas attacks killed over 90,000 soldiers.

Armored Warfare

Armies were quick to realize the advantages of the internal-combustion engine for transporting weapons. By 1910, automobiles were being used as troop transporters and mobile machine-gun carriers.

The idea of a tracked, armored vehicle had been thought of by Italian artist and inventor Leonardo da Vinci (1452–1519). In 1915 British engineers Ernest Swinton (1868–1951) and William Tritton (1875–1956) built the first

The Mark 1 was the first tank used in combat. It was introduced in 1916 and designed to be a movable gun platform that could drive almost anywhere on the war-torn battlefields.

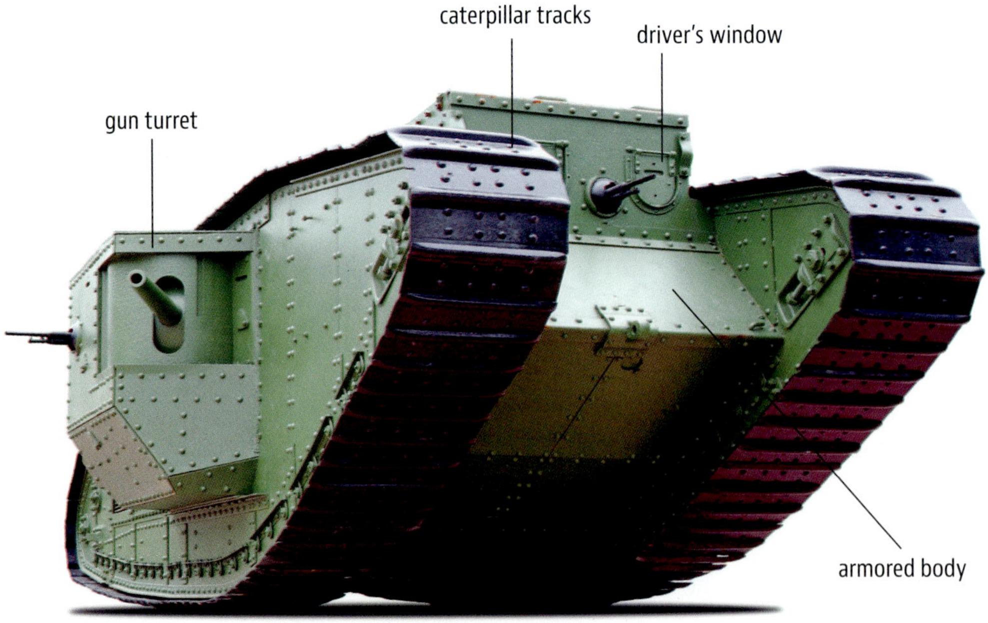

practical vehicle. It was codenamed "Water Tanks for Russia". Tanks combine the technology of caterpillar tracks to cross any terrain; high quality steel armor; modern engines to drive them; and quick-firing guns.

Although the tanks used in World War I were heavy and slow, they could crush barbed wire and help infantry attack enemy trenches. In World War II fast, well-armed tanks (supported by close-support aircraft) were used by Germans forces in a new, fast-moving form of warfare called *Blitzkrieg*.

By 1945, most of the major developments in the infantry weapons, artillery, and tanks that we see today had been reached. However, with the invention of the atomic bomb by the Allies, and guided rockets by the Germans, warfare was about to enter an entirely new and potentially even more destructive era.

TANK FACTS AND FIGURES

- The British Mark 1 tank was armed with six-pounder guns and machine guns. It weighed 31 tons (28 metric tons) and had a top speed of 4 mph (6 km/h).
- The Russian T-34 tank was produced in vast numbers during WWII. Simple and robust, it weighed 39 tons (36 metric tons) and later versions carried a 3.3 inch (85 mm) gun.
- The U.S. Sherman tank was the main tank used by the Western Allies during WWII. It weighed 36 tons (33 metric tons) and was armed with a 3 inch (75 mm) gun.
- The German King Tiger tank was the most powerful armored fighting vehicle of WWII. Armed with a 3.5 inch (88 mm) antitank gun, it had 9 inch (230 mm) thick steel armor and weighed 76 tons (69 metric tons).

THE HOLOCAUST

While chemical weapons were not used in battle during WWII, Nazi Germany employed poisonous gases in Adolf Hitler's "Final Solution"—the extermination of the "undesirable people." Throughout the German-occupied areas of Europe, Jews, Roma (gypsies), homosexuals, and disabled people were systematically killed. Some became slaves in German factories until they perished from exhaustion or starvation, and many more were sent to death camps, such as Auschwitz in Poland. They were killed in large groups inside gas chambers. Called the Holocaust, this persecution claimed the lives of at least 11 million people, six million of whom are estimated to have been Jewish.

The gate at Auschwitz welcomed inmates with the German words for "work makes you free." However, Auschwitz was not a work camp. It was designed to exterminate more than 1,000 people every day.

The first hydrogen bomb was tested on the Pacific atoll of Enewetak in 1952. The explosion produced the mushroom cloud that is typical of nuclear bombs.

The nuclear bomb, a new weapon of devastating power brought World War II, to an end. Since then, scientists have developed military technologies that can destroy any point on Earth.

In early August 1945 atomic bombs were dropped on the Japanese cities of Hiroshima and Nagasaki, causing death and destruction on an immense scale. These events led to Japan's surrender and the end of World War II (1939–1945).

The groundwork for the development of nuclear weapons was laid by British physicist Ernest Rutherford (1871–1937), who discovered that atoms, which are the building blocks of all matter, actually consisted of smaller particles. Rutherford also demonstrated that the atoms of some elements could be "split" to form other elements.

A use for these discoveries become apparent in 1939, when German chemists, Otto Hahn (1879–1968) and Fritz Strassmann (1902–1980), succeeded in splitting the uranium atom, releasing a vast amount of energy, as had been predicted by German physicist Albert Einstein (1879–1955). If a reaction of this type could be performed on a larger scale, it would produce an immensely powerful explosion.

Research into nuclear weapons was not a high priority in Nazi Germany, but the U.S. Manhattan Project, led by J. Robert Oppenheimer (1904–1967) and involving many of the pioneers of nuclear physics from the U.S. and Europe, produced the first atomic bomb in July 1945. At first, only the United

H-BOMB WARHEAD

This is one of several designs of a thermonuclear bomb, which uses heat and nuclear reactions to create an explosion. The fission device is a small nuclear bomb that explodes creating huge temperatures. This causes the lithium deuteride to release radioactive hydrogen atoms, which fuse with each other releasing an immense amount of energy. This second explosion then causes a third explosion in the uranium jacket.

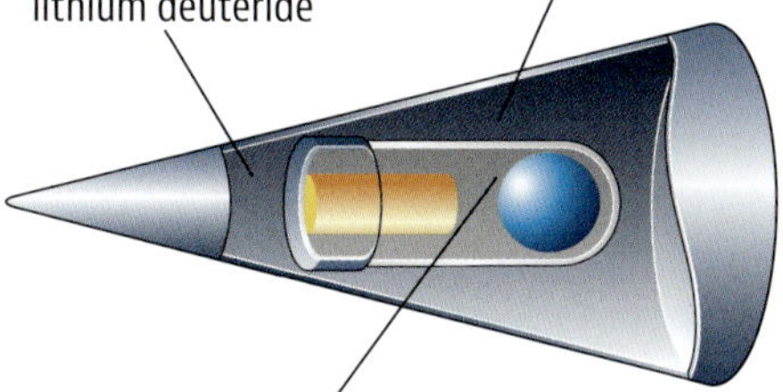

States possessed the atomic bomb. Then in 1949 the Soviet Union tested an atomic bomb.

The Hydrogen Bomb

In 1941, two American physicists, Enrico Fermi (1901–1954) and Edward Teller (1908–2003) were working on the Manhattan Project. They saw that it would be possible to use the energy from an atomic bomb to trigger an even more powerful nuclear reaction in which atoms of deuterium, a rare form of hydrogen, are fused together. In 1949, when the Soviet Union developed an atomic bomb, the United States decided to build one of these "hydrogen (sometimes called thermonuclear) bombs." When detonated, a single hydrogen bomb could produce an explosion equivalent to millions of tons of TNT.

Computers in Warfare

The first computer used in war was developed in the 1930s by Polish scientists who used it to break German military Enigma codes. Influenced by the success of this machine,

MANHATTAN PROJECT BOMBS

"LITTLE BOY" URANIUM BOMB

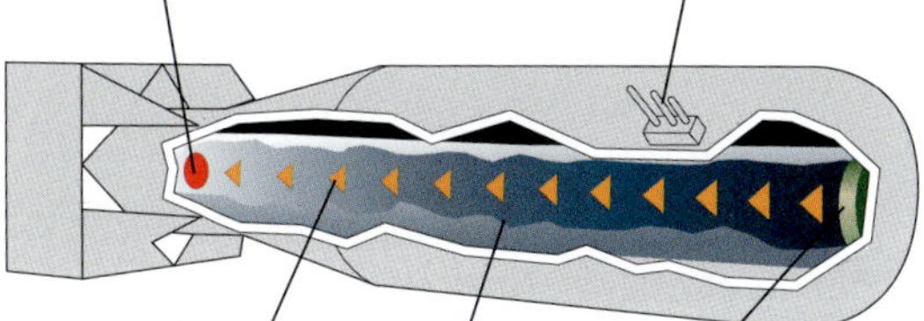

"FAT MAN" PLUTONIUM BOMB

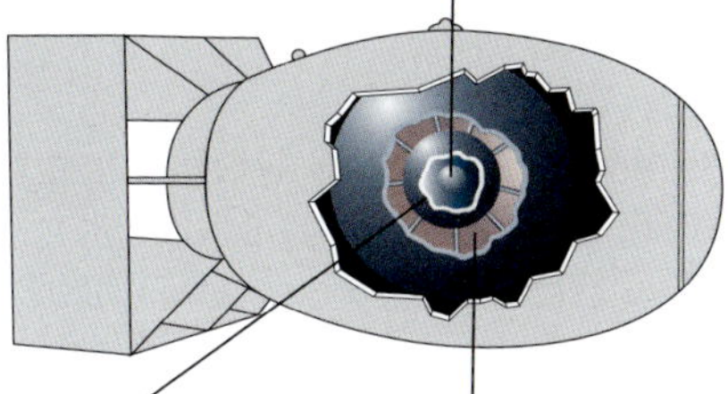

When a message was typed on the Enigma machine's keyboard, each letter was converted into another, shown by lights on the upper set of letters. The code was controlled by a series of adjustable cogs inside the machine, which could be set to more than 17,000 different combinations.

a team of British scientists and engineers led by Alan Turing (1912–1954) built a far more powerful computer in 1943. This computer, named Colossus, was highly successful at decoding German messages during WWII. After the war, a number of large military computers were constructed. Although there were great improvements in processing power and memory, these machines were heavy, bulky, and unreliable. Computers got smaller and smaller in the 1960s and 1970s. This allowed their use in a wide variety of military roles where rapid and accurate calculations are needed. Computers are now vital components of guided missiles, radar, artillery targeting, and most importantly battlefield communications.

Communications

In both World War I (1914–1918) and World War II (1939–1945) communications were conducted by telegraph, telephone, and radio. During battles, both sides were often able to listen in on each other's communications, which were

TACTICAL NUCLEAR WEAPONS

Although the power of the first nuclear weapons was almost limitless, their usefulness on the battlefield was limited by the fact that they were so large and heavy. During the 1950s, however, nuclear weapons that were small enough to be fired from artillery pieces or carried by fighter aircraft were developed by both the United States and the Soviet Union. These weapons, designed to target enemy troops or tanks rather than large civilian populations, are called tactical nuclear weapons.

The B-1 Lancer bomber was designed for the U.S. Air Force to carry nuclear weapons deep into enemy territory, flying at low altitudes to avoid enemy radar systems.

limited to short transmissions. Commanders might only have a vague idea of where their forces were during a battle.

Small, powerful computers and digital transmissions revolutionized communications in war. The volume of information that could be transmitted, received, and processed was dramatically increased. The location and status of every unit could be relayed by satellite to commanders who were hundreds of miles away.

"STAR WARS"

Efforts were made by both the United States and the Soviet Union to develop an antimissile defense capable of stopping nuclear attacks. In 1983, U.S. President Ronald Reagan announced plans to build a space-based antimissile defense system called the Strategic Defense Initiative (SDI), or "Star Wars." This suggested that laser-armed satellites could be used to destroy incoming missiles before they had a chance to release their warheads. The main problem with the idea was the computing power needed to manage such a battle in space. It was beyond the capabilities of existing or even future developments. It was also estimated that Star Wars would have cost about one trillion dollars.

SOCIETY AND INVENTIONS

Nuclear Weapons

The existence of nuclear weapons created unease in the societies of the industrialized West. If both the Soviet Union and the United States launched their entire nuclear arsenals would anyone survive? There were also fears that a nuclear war might be started by an accident caused by human or computer error.

Within Europe, fears about nuclear weapons were mixed with a suspicion that the United States was planning to fight a nuclear war in Europe. These fears were increased in 1977, when the United States was debating whether or not to deploy the neutron bomb. This weapon was designed to kill through increased levels of radiation while reducing the blast and fallout created by an explosion. In essence, the neutron bomb was designed to kill people without damaging property. Public hostility to this weapon was an important factor in it not being deployed.

In the 1980s, fear of nuclear war increased farther as scientists speculated about possible environmental effects of a nuclear war. They feared huge clouds of dust would be blasted into the atmosphere, blocking out the sunlight and causing a long nuclear winter that would most likely kill all animal and plant life.

Modern militaries rely on communications, and so blocking an enemy's communications is a useful weapon. This is called electronic warfare and involves jamming the enemy's radar and radio links through the transmission of electronic signals. Also of importance is the use of surveillance systems to monitor all radio bands to collect intelligence on enemy communications and the location of forces. To counter these technologies, modern military telecommunication engineers devised computer-encrypted messages that are difficult to decode. These are now common in all modern communications systems.

Early warning radars contained within domes scan the sky in all directions for large flying objects.

MILITARY SATELLITES

The military potential of satellites was recognized in the United States as early as 1946, when the U.S. Navy discussed the possibility of using satellites to control guided missiles. However, the deployment of satellites had to wait until the mid-1950s, when more advanced rockets and electronics became available. The first U.S. military satellite, Samos II, was launched in January 1961 and provided photographs of Soviet nuclear missile sites.

Throughout the 1960s and 1970s an increasing number of military "spy" satellites were launched by the United States and the Soviet Union. The quality of pictures taken from these satellites was so good that it was possible to see even minute details. Modern military satellites are capable of sending back instantaneous live-action video images.

Military satellites are also useful for navigation. The U.S. Global Positioning System (GPS) consists of a network of satellites that continuously relay the time and their locations back to Earth. This system was rapidly extended for civilian use, and now billions of people across the world rely on it to navigate their cars or find where they are using their cellphones.

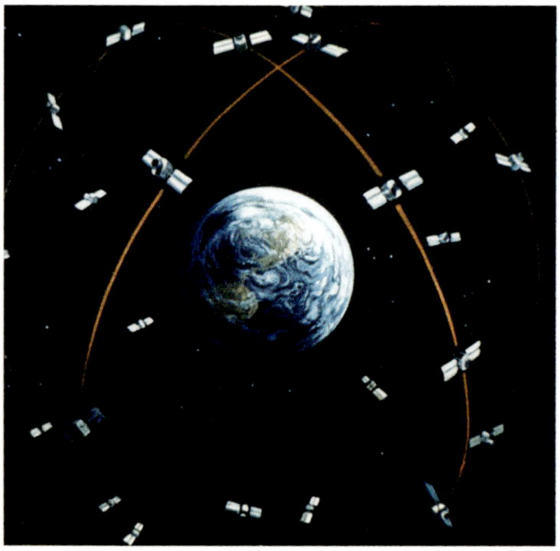

GPS has a network of satellites that circle Earth and send signals to receivers on Earth's surface.

Guided Missiles

By the end of World War II the Germans had made significant advances in the development of guided missiles. The V-2 rocket was able to reach targets up to 200 miles (320 km) away, and several surface-to-air missiles designed to shoot down enemy aircraft had been tested. Both the United States and the Soviet Union made use of captured German scientists and German technology to build guided missiles of their own after the war. Apart from their military use, rocket technology was also used to explore space.

While steady improvements were made to the design and efficiency of rocket engines, it was the advances in electronics, sensors, and computers that allowed guided missiles to become dominant weapons on the modern battlefield, giving rise to a generation of cruise missiles.

Surface-to-air Missiles

In the United States, attention focused initially on developing a surface-to-air missile (SAM) capable of destroying enemy aircraft. By 1953, the U.S. had developed its first SAM, the Nike Ajax air defense missile. The missile was large, cumbersome, and could only be fired from fixed installations. A powerful boost motor accelerated the missile to speeds faster than sound, and a smaller sustainer

CRUISE MISSILES

Cruise missiles are essentially uncrewed high-explosive aircraft. Their engines, which are often air-breathing jets rather than true rockets, burn for a prolonged period (often several hours), and they have short wings that allow them to fly. The German V-1 was the first "cruise missile." It had a speed of 350 mph (560 km/h), a range of 160 miles (260 km), and flew at 2,000–3,000 ft (650–1,000 m).

The V-1 had a small jet engine, which took in air at the front, mixed it with fuel, and ignited the mixture. The resulting hot exhaust gases escaped from the rear, propelling the missile forward. The missile was steered using elevators and a rudder at the rear. Guidance was by autopilot—a compass linked to a gyroscope that was preset on the V-1's target before launch. The V-1 flew a preset distance and was then was put into a steep dive by lowering flaps on the tail section; the engine then cut out and the missile fell to its target.

The United States experimented with cruise missiles in the 1950s but found it impossible to develop an accurate guidance system. By the 1970s, the miniaturization of computers and electronics allowed the creation of new and more accurate guidance systems, which were incorporated in modern cruise missiles. The US Tomahawk cruise missile uses GPS to navigate.

A cruise missile with a warhead fitted

Anti-aircraft missiles are targeted by radar and controlled by computers. When an enemy aircraft is detected, the rocket launcher swivels toward it and raises the missiles to the right angle before firing.

rocket powered the missile over the rest of its flight. Once airborne, radar beams tracked both the missile and the target, sending information to a computer, which guided the missile onto a collision course. One year later the Soviet Union deployed its own SAM, the SA-1, which was guided by a radar located inside the missile itself—a more versatile system than that used in Nike Ajax. In the following decades both the United States and the Soviet Union developed SAMs that were smaller, more agile, and able to engage targets at greater distances.

Ballistic Missiles

Until the 1950s, the United States and the Soviet Union both relied on large strategic bombers to attack each other's homelands in the event of a war. However, the development of SAMs reduced the chances of bombers penetrating the enemy's missile defense systems. In response to this dilemma both countries developed intercontinental ballistic missiles (ICBMs) armed with nuclear warheads.

CHEMICAL AND BIOLOGICAL WARFARE

Poisonous gases were used as weapons in Word War I. During the 1930s, the Germans developed far more deadly chemical agents. Among these were two nerve gases (poisons that kill by disrupting the victim's nervous system) called sarin and tabun, but they were never used during World War II. After the war, the United States, Britain, and the Soviet Union made use of German scientists and captured stockpiles of sarin and tabun to make their own chemical weapons.

Chemical weapons were made illegal in 1968, although they have been used since in a few wars. Biological warfare uses bacteria and viruses to cause disease in people, animals, or plants. Developments in microbiology have made it possible to "weaponize" many diseases, including anthrax and lassa fever. As of 2025 there is no proof that biological weapons have actually been used in war.

Tiny amounts of biological weapons could kill enormous numbers of people if they are ever used.

A Minuteman II ICBM hidden in a deep underground silo. The missile had three rocket stages that powered it into space and over a target in three minutes.

These are multistage rockets similar to those used to put satellites into orbit. They are fired into space. From there the warhead follows an arc, returning to Earth thousands of miles from the launch site. They use a combination of inertial guidance (using onboard motion sensors), stellar navigation (checking the position by reference to the stars), radar, and GPS to hit targets.

The Soviet SS-9 was the first operational ICBM to incorporate multiple warheads, or Multiple Reentry Vehicles (MRVs). The missile

TRIDENT II MISSILE

This model of ICBM is carried on American and British nuclear submarines. They are launched under water and can carry up to 12 warheads for more than 8,200 miles (12,00 km).

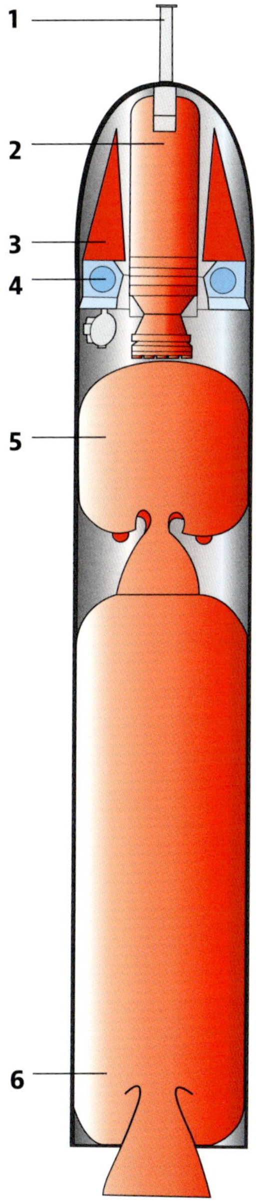

1 The aerospike reduces drag as the missile flies through air.

2 The third-stage rocket motor positions the warheads above their target while in space.

3 The missile can carry 12 MIRV warheads.

4 The equipment section contains the guidance and targeting systems.

5 The second stage rocket motor provides an extra boost to get the missile into space.

6 The first stage motor is used to launch the missile. Like the other stages, this motor is powered by solid rocket fuel.

TYPICAL MISSILE GUIDANCE SYSTEMS

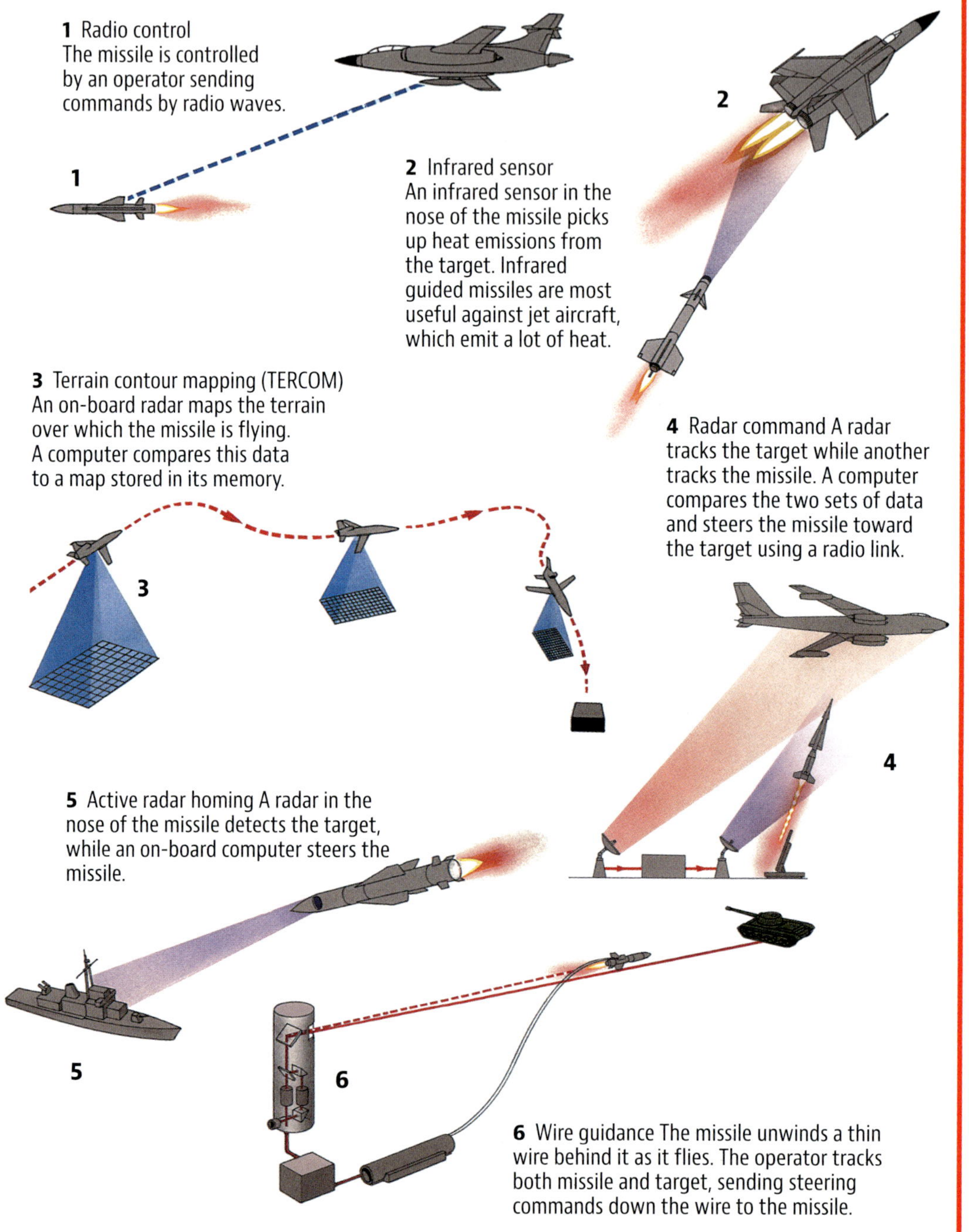

1 Radio control
The missile is controlled by an operator sending commands by radio waves.

2 Infrared sensor
An infrared sensor in the nose of the missile picks up heat emissions from the target. Infrared guided missiles are most useful against jet aircraft, which emit a lot of heat.

3 Terrain contour mapping (TERCOM)
An on-board radar maps the terrain over which the missile is flying. A computer compares this data to a map stored in its memory.

4 Radar command A radar tracks the target while another tracks the missile. A computer compares the two sets of data and steers the missile toward the target using a radio link.

5 Active radar homing A radar in the nose of the missile detects the target, while an on-board computer steers the missile.

6 Wire guidance The missile unwinds a thin wire behind it as it flies. The operator tracks both missile and target, sending steering commands down the wire to the missile.

Infantry soldiers use a computer to control a small reconnaissance drone that can show them enemy activity over the mountain.

carried several smaller warheads that separated from one another during the middle part of the missile's flight. This technology was adopted and developed in the U.S. Minuteman program. Minuteman I, II, and III missiles were in service in 1962, 1965, and 1970. This weapon carried Multiple Independently Targetable Reentry Vehicles (MIRVs), warheads that could maneuver in space, and fly toward a different target.

Antitank Missiles

To pierce tank armor, antitank missiles use shaped-charge warheads, invented in the 1930s. A cone-shaped explosive is detonated inside a thin metal case, creating a jet of extremely hot gas that can punch through thick steel plate. Some antitank weapons use super-dense uranium shells to punch through thick armor.

DRONE WARFARE AND COMPUTING POWER

One of the major developments in warfare since 2000 has been the use of drones. Drones are pilotless aircraft that are controlled byan operator who may be based thousands of miles away. Drones combine GPS technology (which the drone uses to calculate its position) with the power of modern computers (to manipulate the flight of the drone with a high degree of precision) and the efficiency of modern radio communications (to maintain contact between operator and drone). There are many small drones that can be used by soldiers on the battlefield, either for short-range reconnaissance or to attack the enemy. Some drones, known as "kamikaze drones," destroy themselves when they crash onto a target. Drones became important in the warfare in Ukraine from 2022. Ukrainian drones halted invading Russian armored columns by dropping explosives onto the upper parts of tanks.

A US Reaper drone, a remotely controlled aircraft that has been used to target enemy commanders.

SEA WARFARE

Laser guns are efficient weapons. This U.S. Navy AN/SEQ-3 Laser Weapon System can be programmed to disable an enemy vessel or to destroy it completely.

Oceans cover two-thirds of the world's surface. They have played an important role in human history, from trade to exploration. Not surprisingly, seas have also been a key area for warfare.

The first warships were galleys, a design that was so effective it was used for 2,500 years. Galleys were powered by a combination of sails and rows of oars arranged one above the other in banks. In battle the oars made the galley maneuverable and provided the power needed to thrust the ship's main weapon, a ram, into the enemy ship's hull to disable it.

In ancient times galleys were named for the number of oars they carried, or the

GREEK TRIREME

The invention of the trireme, with its three sets of oars, around 500 BCE transformed the Greeks into a great naval nation. Triremes were usually around 150 ft (46 m) long and carried a crew of 200, of whom 170 were rowers grouped in tiers of three. The oarsmen were highly trained crew members. Unlike later vessels, triremes only employed one person per oar. Through this arrangement, the builders of the triremes could fit a large number of rowers in a relatively small space, making the boats easy to maneuver and able to accelerate quickly.

Triremes were formidable opponents. They could cruise at 6 knots (7 mph, or 11 km/h) for long periods of time and accelerate to 12 knots (14 mph, or 22 km/h) when ramming. The trireme has been described as a ram with a ship attached. It was designed to fit the maximum number of rowers (to propel the craft with great force) in the smallest amount of space, allowed by its ramlike shape. However, the long, thin, top-heavy design made the trireme unstable, so its hull was weighed down with stones. Leather shields protected the rowers from enemy missiles.

As this computer-generated illustration shows, triremes relied on oars during battle but used sail power for long journeys. The ship was steered by large oars, or steerboards, attached at the rear of the ship.

COPPER-BOTTOMED SHIPS

One of the most significant pieces of metal and ship technology to develop in the 18th century was the technique of covering ships' hulls below the water line with copper sheets. The aim of this was to correct a serious problem faced by all wooden sailing ships—particularly those sailing in the warm waters of the Caribbean and Southeast Asia—the teredo, or shipworm. This creature would eat its way through the hull of ships, shortening the life of their timbers. After several ineffective solutions were tried, the idea of using copper to plate boat's hulls was put to the British Royal Navy in 1708. The navy rejected it on the grounds of cost, but by 1761 there was an urgent need to find a solution to the problem. HMS *Alarm* became the first warship to have a copper bottom. The technique was soon in common use on both naval and merchant (commercial) ships.

The copper plate on the hulls of ships often went green from corrosion by the salty sea water. Iron and zinc was added to the copper to reduce this problem.

HMS Victory, *the flagship of the British navy in the early 19th century, had 104 cannons, mostly firing through gunports along the sides of the vessel.*

number of oarsmen. The first war galleys, built and operated by the Phoenicians (from Lebanon) between 1100 and 800 BCE, were called penteconters. These small galleys had a single row of 50 oarsmen. They were 50 ft (15 m) long and could sustain a speed of 6 knots (7 mph, or 11 km/h) for a short time.

Penteconters were soon replaced by faster Greek galleys, such as the trireme, with its three banks of oars, and the larger and even faster septireme, which assigned three or four oarsman to each oar. The design of the galley was developed further still by the Romans, who equipped their ships with catapults and a large plank with a spike on the end called a corvus. This was dropped onto the enemy ship to secure it, while marines boarded and captured the vessel.

The Age of Gunpowder

Although galleys continued to be used in warfare by the Swedes and Russians until as late as 1790, the widespread use of gunpowder weapons on ships from the 16th century marked the point at which they went into decline. In a world where an opponent could bombard the enemy with guns from a distance, galleys and their close-quarter tactics had no future.

The first true sailing warship was the cog, built in the 12th century. This was the result of two inventions. First came the introduction of the stern rudder, which allowed the ship to be steered effectively in difficult winds. Second, ships were built with deeper and stronger hulls that made them steadier in rougher seas. In the early 16th century, heavily armed galleons came to dominate naval warfare. In 1501, the French invented the gun port.

CONGREVE ROCKETS

The idea of cities being bombarded by rockets may seem a modern one. However, in 1805 the British navy began using Congreve rockets to attack ports. They could start fires and fly over walls and other defenses. The rockets were developed by Sir William Congreve (1772–1828), an army officer who had seen the weapon in action when fighting rocket-armed forces in India in the 1790s. Encased in metal, Congreve's rockets were up to 3 ft (1 m) long, and weighed around 32 lb (15 kg).

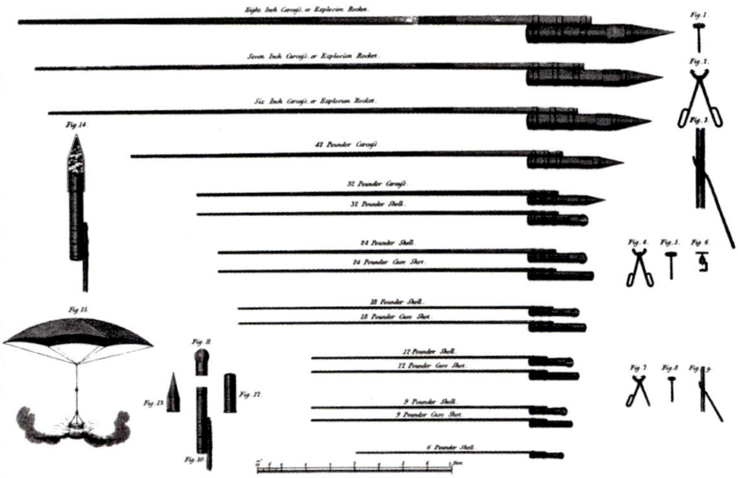

A range of Congreve rockets weapons were developed. To improve accuracy, the rockets were attached to long sticks. Some were 15 ft (4.5 m) long.

This was a hatch or windowlike opening in the side of the ship that could be opened in order to fire the guns and then closed in bad weather. This allowed guns to be placed below decks. The result was that a far greater number of guns could be placed aboard sailing ships. English sailor Sir John Hawkins (1532–1595) removed the high front and back of the galleon to create a steadier vessel, later called a ship-of-the-line, which could carry several layers of cannon on its decks. These ships did maximum damage when all the guns on one side were fired together (called a broadside). This design remained the basis for warship construction until the mid 1800s.

The iron-hulled steamship HMS Warrior, *seen here with a wooden-hulled ship-of-the-line (left) in 1864, used sails for auxiliary power.*

Steam and Steel

Industrial development in the 19th century inevitably revolutionized warship design. In 1783, French engineer the Marquis de Jouffroy (1751–1832) built the steam-propelled *Pyroscape*, which was driven by a paddle at the rear of the ship. Other designers copied his idea, but these ships were unsuited to the rough open seas. The first steam warship, the *Demologus*, was designed in 1813 by the U.S. engineer Robert Fulton (1765–1815). However, the paddle used to propel this ship was vulnerable to enemy gunfire. Later, warships were powered by screw propellers. Steamships continued to use sails as an additional source of power until the 1870s.

Another impact on warfare at seas was the invention of warships with iron and, later, steel hulls. The idea of using iron plates to protect the decks of wooden hulls had been used as early as the 1590s by Korean warship designers, but in 1860 the British took the idea further. HMS *Warrior* became the first warship with a hull made solely of iron. *Warrior* was 480 ft (145 m) long, carried 48 smoothbore guns on its gun deck and had a speed of 14 knots (17 mph, or 26 km/h).

In 1862, U.S. shipwright John Ericsson (1803–1889) made a key contribution to the development of warships. With the launch of *Monitor* he radically changed the way guns were placed on warships. The vessel's guns were positioned in a rotating turret on the deck of the ship, which allowed them to fire at targets to the front and rear of the ship, not just to the side.

In 1871, the British naval architect Edward Reed (1830–1906) combined a steam engine, a steel hull, and the massive firepower of 12.5 inch (30 cm) guns mounted in turrets in the 14,000 ton (12,700 metric ton) HMS *Devastation*—the first truly modern

GLOIRE

In 1859 the French built *Gloire*, which was the first warship with a wooden hull covered in iron. Such ships were called ironclads.

Designed by Dupuy de Lôme, the iron cladding was added as protective armor for the French frigate. This covering stretched from a point 6 feet (nearly 2 m) below the waterline to the upper deck, and was 4.75 inches (12 cm) thick. The layer was completed by a 26 inch (66 cm) oak backing.

THE SCREW PROPELLER

Although John Ericsson had invented the screw propeller in 1836 and fitted it the to U.S. warship *Princeton* in 1843, navies around the world were reluctant to adopt it. It was argued that although it was not as vulnerable as the paddle, it lacked power. The argument was finally resolved in a dramatic manner in 1845. The British navy lashed the paddle wheeler *Alecto* and the screw propeller-driven *Rattler* together in order to see which was the stronger. On April 3 the contest began in the North Sea. Initially, there was little movement as each vessel increased power. Then, slowly, the *Rattler* inched forward, gathering speed until it was towing the 900 ton (815 metric ton) *Alecto* backward. The contest proved the power and superiority of the screw propeller.

The brass screw propeller proved itself far superior to the paddle wheel during Royal Navy tests in 1845.

battleship. Battleships continued to grow in the early 20th century, the largest being the Japanese warship *Yamato*. However, during World War II (1939-1945) these huge warships proved vulnerable to a new technology: aircraft carrying bombs or torpedoes that could be launched from an aircraft carrier at sea.

Era of the Aircraft Carrier

The first time an airplane was flown from a ship was November 14, 1910, when Eugene B. Ely took off from the U.S. cruiser *Birmingham*. The following year Ely became the first person to land an aircraft on a ship. The first ships used as aircraft carriers were simply modified

SOCIETY AND INVENTIONS

Clash of the Ironclads

In 1862, two ironclad warships went head to head for the first time at the Battle of Hampton Roads, taking place along a wide waterway in Virginia. The battle was the most important naval engagement of the U.S. Civil War. The Confederate ship *Virginia* was armed with 12 guns, which were fired through portholes positioned along the ship. It had proved itself superior to any Union wooden-hulled vessels. The Union vessel *Monitor* had just two guns, but they were encased in a turret that swiveled side to side. The battle lasted two days and ended as a draw. Neither ship could sink the other.

Despite being out-gunned, Monitor *(center) matched* Virginia *(left) for firepower thanks to its rotating gun turret. Within decades, all battleships used similar turrets. Aiming guns accurately from these turrets became a major part of naval training.*

DREADNOUGHTS

The introduction of HMS *Dreadnought* in 1906 was a turning point in the history of battleship design. The impact of the vessel was so great that battleship design is divided into two eras: pre- and post-dreadnought. Not only was it the first warship to have steam turbine propulsion, it was also the first "all-big-gun" ship in the world. Previously, battleships had tended to be armed with both a main battery of high-caliber guns and supporting batteries of smaller weapons. *Dreadnought*'s main armament was all of one caliber, consisting of ten 12-inch (305 mm) guns in five turrets. The 1914 Dreadnought class battleship shown here, the USS *Texas* (now a museum ship near Houston) is a super-dreadnought. It carried ten 14 inch (360 mm) caliber guns housed in five twin turrets. *Texas* also had 21 smaller 5 inch (130 mm) guns.

warships with a flat deck at the rear. HMS *Argus,* launched in 1918, had a full unobstructed flight deck and was the first modern aircraft carrier. In subsequent years the British Royal Navy led the world in the technical design of carriers.

To assist with landing the aircraft, rows of wires were placed across the landing deck to snag hooks that trailed from the aircraft. These stopped them rolling off the deck into the sea. In 1920, *Eagle* was the first ship to have its bridge and funnels set to one side of the flight deck in order to create more room for landing. In 1951, the use of heavier jet fighters led the British to invent a steam catapult, which is used to assist the jet as it takes off by giving an extra boost of power. In 1961, the U.S. Army launched the first supercarrier—the 84,000 ton (76,000 metric ton) *Enterprise*. The latest supercarriers of the U.S. Gerald R. Ford class launch aircraft electromagnetically, dispensing with the need for steam catapults.

FACTS AND FIGURES

- At over 80,000 tons (72,000 metric tons), the Japanese *Yamato* was the heaviest battleship ever seen. The WWII warship was 862 ft (263 m) long and had a crew of 2,500. Among its array of weapons were nine 18 inch (46 cm), twelve 6 inch (15 cm), and twelve 5 inch (13 cm) guns. Its armor was designed to withstand a 2,200 lb (1,000 kg) bomb dropped from a height of 10,000 ft (3,000 m). *Yamato* was sunk by U.S. airplanes on April 7, 1945.
- The record speed any kind of military vessel has achieved is 91.9 knots (110 mph, or 170 km/h), a record set by the U.S. attack hovercraft SES100B on January 25, 1980.

Underwater Weapons

Technological developments have allowed war to be fought under the sea, as well as on and above it. The American inventor David Bushnell (1742–1824) devised the mine, an underwater explosive, in 1775. The torpedo was the joint invention of the Englishman Robert Whitehead (1823–1905) and the Austrian Giovanni Luppis (1813–1875). In 1866, Whitehead built an underwater missile that was 14 ft (4 m) long and driven by compressed air at a speed of 6 knots (7.2 mph, or 11 km/h) for 700 yards (640 m). Although widely used in World War I (1914–1918), technical problems made torpedoes unreliable. The problems were overcome in the 1930s by the Japanese, who invented the Long Lance torpedo with a speed of 49 knots (56 mph, or 91 km/h) and a range of 11 miles (18 km). Later, torpedoes had electronic guidance systems.

The most dangerous underwater weapon system is the submarine. In 1578, the Englishman William Bourne first put forward the idea, but it was the Dutchman William van Drebble who built the first one on the Thames River in 1620. In 1776 David Bushnell's *American*

THE *AMERICAN TURTLE*

The world's first military submarine was created by David Bushnell. He began work on the design of the vessel while still a freshman at Yale in 1771. The American Revolution (1775–1783) gave Bushnell the opportunity to put his ideas into practice and, in 1775, he created the *American Turtle*. The *Turtle* was a one-person submarine driven by a hand-operated propeller. The operator was also responsible for steering and letting water in and out of the ship's tanks for ballast (weight). The idea was that *Turtle* would creep up under British warships, attach kegs of gunpowder to their hulls, and then retreat to a safe distance to watch the resulting explosion.

Unfortunately for Bushnell, *Turtle* proved unsuccessful in battle: its attack on the British man-of-war HMS *Eagle* failed when the gunpowder could not be attached and the resulting explosion proved incapable of damaging the British warship's hull.

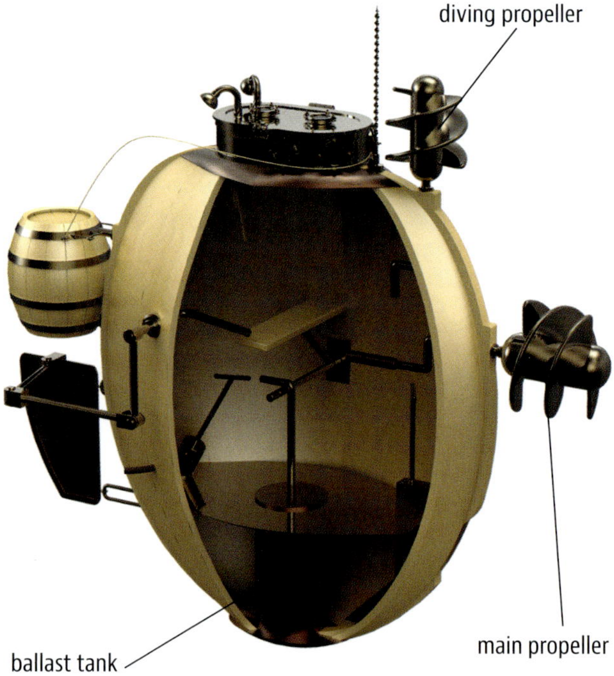

The American Turtle *was a large wooden barrel with a brass viewing compartment at the top. The vessel submerged when water was pumped into the ballast tank. The operator could then go deeper by cranking the diving propeller.*

This World War II submarine has two torpedo tube doors on the bow. The submarine was steered underwater using adjustable diving planes. On the surface, the captain could control the vessel from the conning tower.

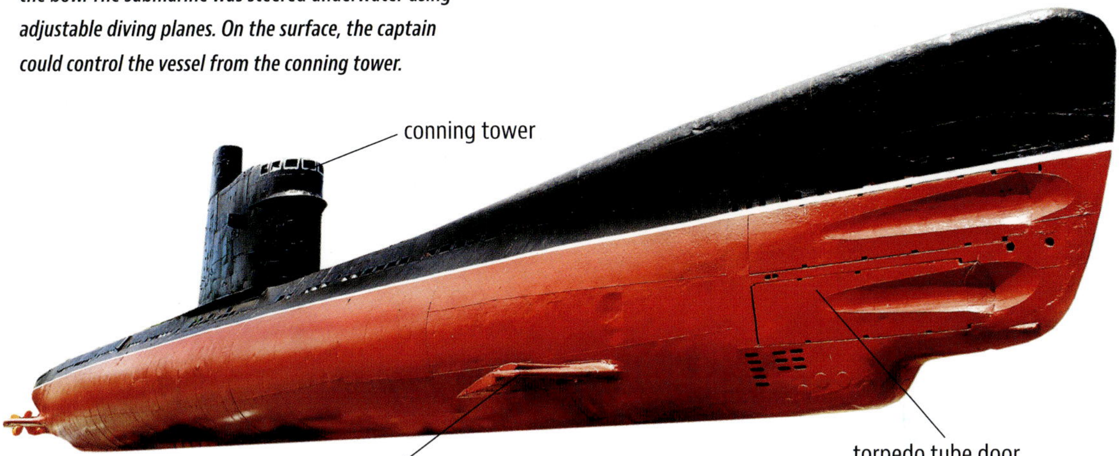

SUPERCARRIER

Aircraft carriers are the largest warships afloat today. The U.S. Navy's Gerald R. Ford class are supercarriers. They have a crew of more than 4,000 and can carry more than 75 aircraft.

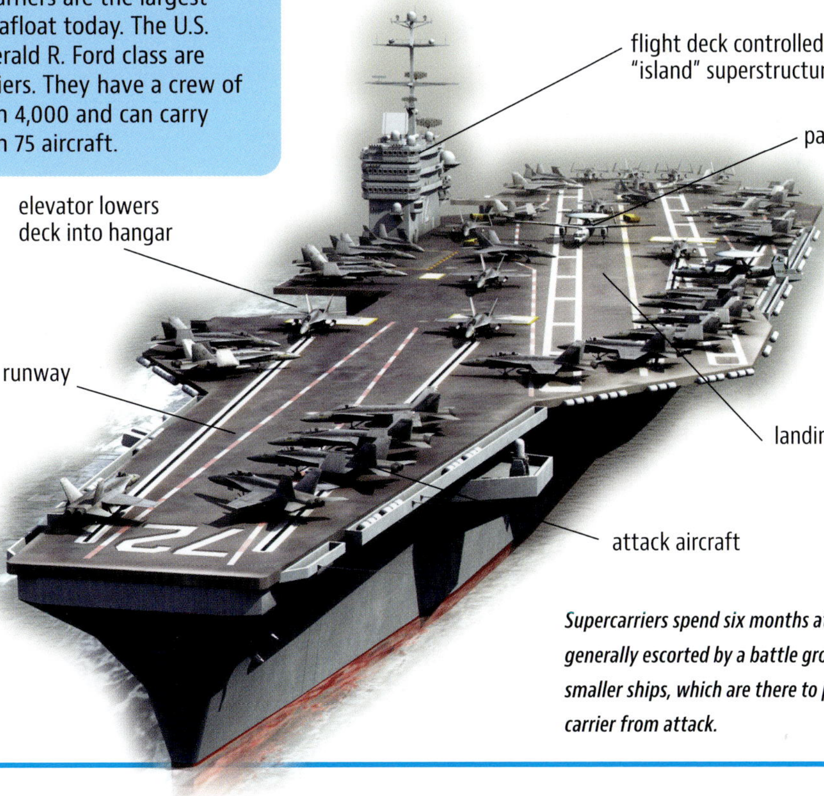

Supercarriers spend six months at sea, generally escorted by a battle group of smaller ships, which are there to protect the carrier from attack.

Torpedoes are fired from watertight tubes. Once loaded, the inner doors are locked and the outer ones opened to flood the tubes, making the torpedo ready to fire.

Turtle went into service as the world's first military submarine.

It was the invention of electric and fuel motors in the 1880s, along with the torpedo, that gave submarines weaponry and an effective means of propulsion. However, these submarines had to rise to the surface of the ocean to recharge the electric batteries used to power the craft underwater. Diesel engines were used to power the submarine at the

NUCLEAR-POWERED SUBMARINES

Nuclear submarines are powered by a pressurized-water reactor (PWR) that enables them to stay underwater indefinitely. In a PWR, a uranium-based nuclear reactor heats a primary circuit of water, which is pressurized to prevent it from boiling. This, in turn, heats a steam generator that marks the start of the secondary circuit. The water in this is transformed into steam, which drives a turbine that is directly connected to the submarine's propeller. The steam then travels through a series of condensers, which turn it back into water. This then flows back into the steam generator, and the whole process starts again.

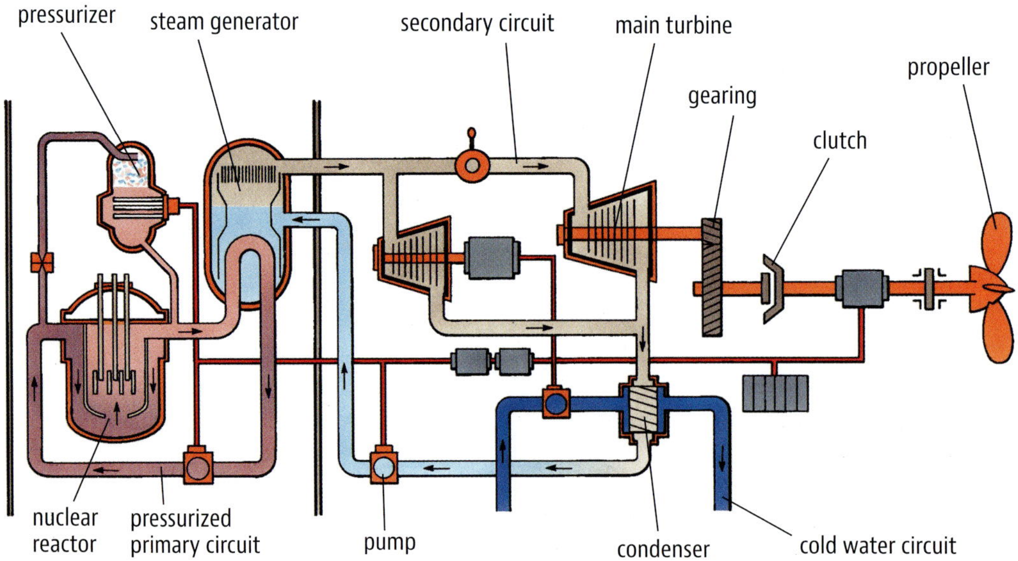

SOCIETY AND INVENTIONS

Submarines and the Nuclear Age

The advent of the nuclear-powered, nuclear-armed submarine in the second half of the 20th century had a vast effect that stretched far beyond the realms of naval warfare. The new nuclear-powered vessels are capable of traveling virtually indefinitely without surfacing—until food supplies run out—and are very difficult to detect. Since they are often armed with nuclear missiles with 2,500-mile (4,000-km) ranges, no target in the world is beyond their range. Submarine-based nuclear arsenals are almost completely beyond attack by enemies and that ensures that if one country launches a nuclear attack, it is inevitable that a devastating strike will be made in retaliation.

NUCLEAR SUBMARINE

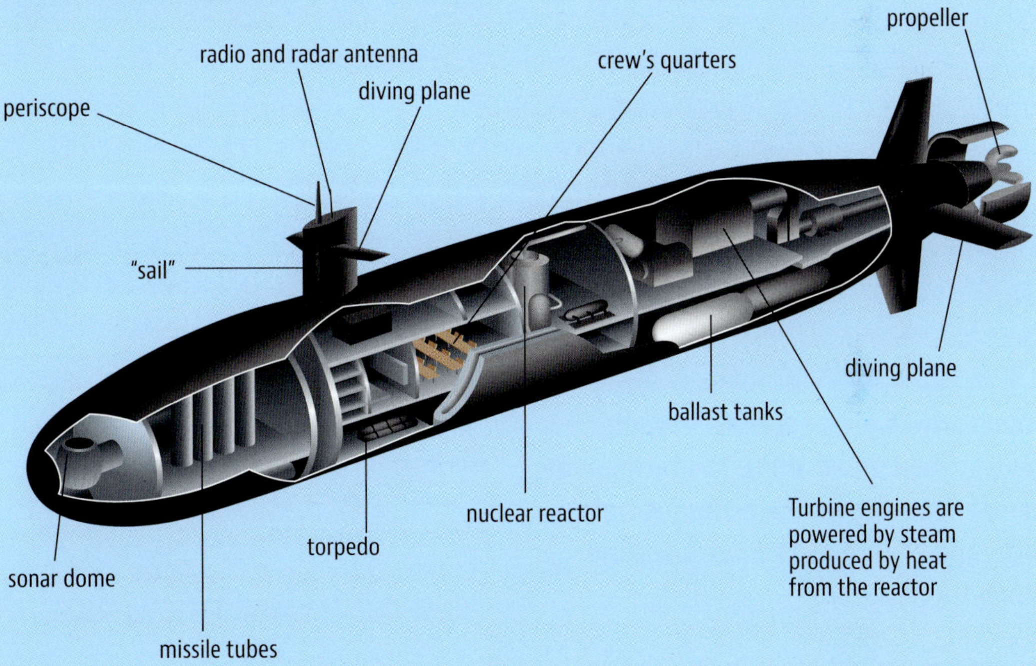

surface at speeds of up to 10 knots (12 mph, or 19 km/h) while recharging their batteries ready to submerge again.

In 1955, *Nautilus*, the first nuclear-powered submarine, was built. The nuclear power plant enabled *Nautilus* to sail right around the world while submerged. The addition of nuclear ballistic missiles in the 1960s, meanwhile, created submarines capable of destroying cities thousands of miles away, making them not only difficult to detect but also a potent threat.

LOCKS AND KEYS

Castles needed strong defenses to keep invaders out. Many castle gateways were secured by a huge iron grill, called a portcullis, that could be raised to let people in, and lowered to keep out uninvited visitors.

From early times, people have wanted to protect their possessions. The technology has developed over the centuries from locks and keys to electronic codes and biometric security.

One of the earliest known locks was found in the ruins of the palace of Khorsabad in Iraq. Made entirely from wood, this lock is probably 4,000 years old. The door is secured with a sliding wooden bolt. The lock mechanism, called a pin-tumbler, consists of several holes drilled through the bolt with matching pins attached to the door above. As the bolt is slid across the door, the pins drop into the holes, locking it in place. The key is a large wooden bar with a group of upright pegs that

FACTS AND FIGURES

- Mortise locks have the locking mechanism mortised (meaning cut) into the door. They are very strong locks when installed in heavy doors.
- Rim locks, including their bolts, are enclosed in a square metal case and mounted on the inside of the door.
- Cylindrical (or knob) locks form part of the doorknob—the keyhole is actually set in the knob of the door handle.

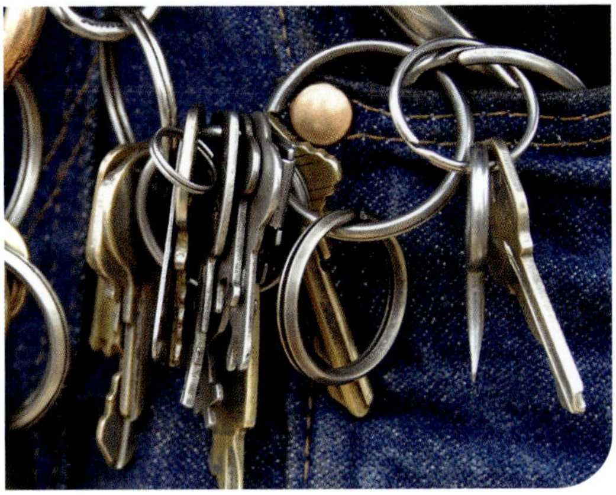

Every lock in a building usually needs a different key, carried together on a key ring. Janitors sometimes have a master (or "skeleton") key that opens every lock in their building.

matches the pattern of holes. When the key is inserted, the pegs lift the pins out of the bolt, and the door is unlocked. Locks of this type are still used in parts of North Africa and the Middle East, and the falling pin design forms the basic principle of many locks, including the famous Yale lock.

Locks in Ancient Greece

The ancient Greeks were the first people to use metal keys, but their locks were less sophisticated than the older pin-tumbler designs. The bolt was moved by an L-shaped key made from iron. The key was passed through a hole in the door and turned, engaging the bolt and drawing it back. This type of lock provided little security because it was not encoded. Anyone who could make an iron L-shape of the right dimensions could open the door.

Locks in Ancient Rome

Roman locks were the first to be made entirely from metal, usually iron for the lock and bronze for the key. The Romans invented wards, which are intricate patterns of matching slots on both key and lock.

HARRY HOUDINI

One of the greatest and most famous lock-pickers of all time was Hungarian-born American illusionist Harry Houdini (1874–1926). Houdini specialized in entertaining thousands of people with his escape acts. In a typical act he would be shackled with metal chains and locked inside a wooden box. The weighted box was then submerged in water, leaving him only a few minutes to break free before he was drowned. In addition to his skill at manipulating locks, Houdini relied on physical agility and trickery to make his escapes.

Harry Houdini prepares to jump from Harvard Bridge, in Boston, while locked in chains.

A simple lock on a medieval church door. The latch is a hook that can be used to open and close the door from both sides, while a large bolt locks the door from the inside only.

Once the key is slotted onto the ward, it can be rotated around the ward rail to operate the bolt. Warded locks have always been relatively easy to pick because special keys could be made to pass around the ward even if they did not fit perfectly.

Locks in the Middle Ages

During the Middle Ages a great deal of skill was used in crafting locks, but the technology remained the same. Keyholes were often hidden behind secret shutters and false keyholes were often cut into the door to confuse thieves. However, security still depended on complex warding. One new innovation was the English letter-lock of the 17th century. A number of rings inscribed with letters or numbers were threaded onto a spindle. When the rings were turned to form a particular number or word, slots inside the rings were aligned and the spindle could then be drawn out.

Early Modern Locks

The 19th century was a time of major advances in lock design. In 1778, English locksmith Robert Barron patented a double-acting tumbler. A tumbler is a lever that falls into a slot and cannot be moved unless it is raised to a set height. Inserting the key raises the tumblers, while turning it slides the bolt out. The Barron lock had two tumblers, and the key had to raise each by a different amount before the bolt would slide. The principle behind this design is used in all lever locks today, but a skilled lock picker could open Barron's lock.

In 1784, an entirely new lock was patented by an English engineer Joseph Bramah (1748–1814). It only had a small key but it provided a level of security never achieved before. Bramah constructed a geared lathe to manufacture his locks. This was one of the first metalworking machines designed for mass production. The locks were very expensive, but Bramah was so sure of their security that he offered a reward of £210,

BRAMAH'S LOCK

Joseph Bramah's 1784 patent lock used a key that had slots of different lengths cut into the end lengthways. When the key was pushed into the lock, it depressed a number of internal mechanisms called slides to specific depths, depending on the length of the slots on the key. Only when all the slides were depressed to the right depth could the key be turned and the bolt drawn. Bramah's mechanism was similar to the D-shaped locks used to secure bicycles.

an enormous sum in those days, to the first person who could pick one. This challenge stood for 67 years until a locksmith in the United States, A.C. Hobbs, succeeded and claimed the reward. However, picking the lock had taken Hobbs an extraordinary amount of time: more than 50 hours!

Yale Locks

By the middle of the 19th century the lock industry was booming. The fast-growing economies of the United States and Europe provided a huge demand for locks, and hundreds of new types of lock were patented. Most of these were merely variations on existing lock designs. In 1848, Linus Yale (1821–1868) patented a pin-tumbler lock based on ancient principles, and in the 1860s he designed the Yale cylinder lock, with its familiar serrated (jagged-edged) key. Pins are raised to the correct heights by the serrated edge, so releasing the cylinder, which can then be turned.

PADLOCKS

Padlocks are removable locks with a hinged or pivoting bar called a shackle, which can be used to join two objects together. The locking mechanism is enclosed in a brass case. To shut the lock, the shackle is pushed into a hole in the case. Inside, a lever falls into a slot at the tip of the shackle, holding it in place. A combination or a key is used to release the shackle and open the lock.

COMBINATION LOCKS

The combination lock, based on 17th-century English letter-locks, consists of a number of rings attached to a central spindle. When the correct combination is entered, the spindle can be drawn out because slots inside the rings are aligned. A combination lock with four rings and 100 numbers on each ring offers a picker 100 million different options. Because there are many possible combinations and no keyholes in which an explosive charge can be placed, the combination lock became the most secure form of protection for safes and strong rooms in the 19th century. An added bonus is that combinations can be easily changed. To try to unlock combination safes, some lock-pickers have used stethoscopes to listen for telltale clicks as the rings inside the combination lock fall into line.

Time Locks

In the 1870s, a new crime wave swept across the United States. Armed robbers would go into banks and force cashiers to open their safes. To combat this, an American locksmith built the first time lock in 1873. The lock contained a built-in clock and could only be opened at a preset time. There was no longer any point in threatening the staff of a bank with violence.

Modern Locks and Keys

The importance of locks as a protection against bank robbers declined after World War II (1939–1945), during which time knowledge of explosives became widespread. Locks became more difficult to pick, and thieves usually ignored them and simply blasted their way into a bank vault or safe.

Modern car keys send a radio signal to the car, which activates the doors' magnetic locks and unlocks the car.

THE GOLD VAULT

The vault beneath the Federal Reserve Bank of New York contains gold reserves worth $86 billion and must be one of the most secure places on Earth. Made from steel-reinforced concrete, the three-story bunker is sunk 80 ft (27 m) below street level, where it rests on the bedrock of Manhattan Island. There are no doors into the gold vault. Entry is through a 10 ft (3 m) passageway cut through a 90 ton (80 metric ton) steel cylinder that revolves vertically in a 140 ton (127 metric ton) steel-and-concrete frame. To seal the vault, the cylinder, which is slightly tapered, is rotated through 90° and lowered into the frame. Eight large bolts are then inserted into the cylinder to secure it. Bank personnel can open the vault by unlocking a series of time and combination locks, but no one person has all the necessary combinations. In addition, the vault has armed guards and an electronic surveillance system. No one has ever tried to steal the gold.

The walls around a bank vault or strongroom are too thick to cut through or blow a hole in.

TYPES OF LOCKS

1 A warded lock relies on a specific matching pattern of slots and grooves on both lock and key. Once inserted, the correct key can be slid onto the ward rail and rotated, drawing back the bolt.

2 A Yale cylinder lock uses the pin-tumbler principle. The pins are cut into two parts of varying lengths. Notches cut into the flat key raise each pin to a specific height at which the cut between the two parts of the pin aligns with the edge of the cylinder. The cylinder can be rotated, releasing the locking bolt.

3 A combination lock consists of a series of rings threaded onto a central spindle. A slot is cut into the outer edge of each ring, and a raised stud is present on the inner surface. When the outer knob is rotated, the pin engages an arm on the spindle, which in turn engages the stud on the first ring. The rings are connected to one another by the same mechanism. By turning the knob through a specific series of clockwise and counterclockwise rotations, using the numbers etched onto the dial surface as a reference, the slots on the rings are aligned, allowing the bolt to be drawn out.

1

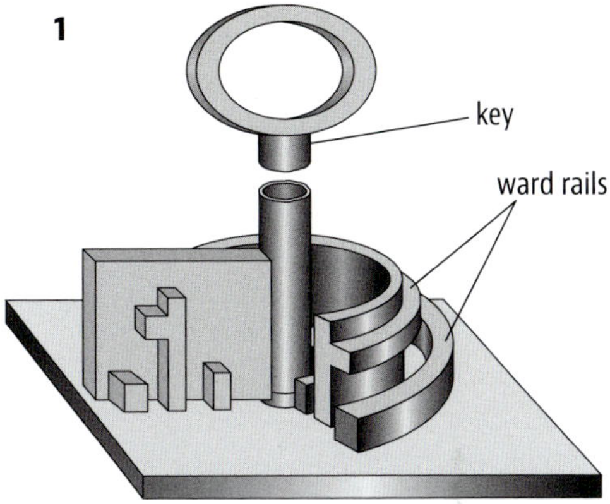

2

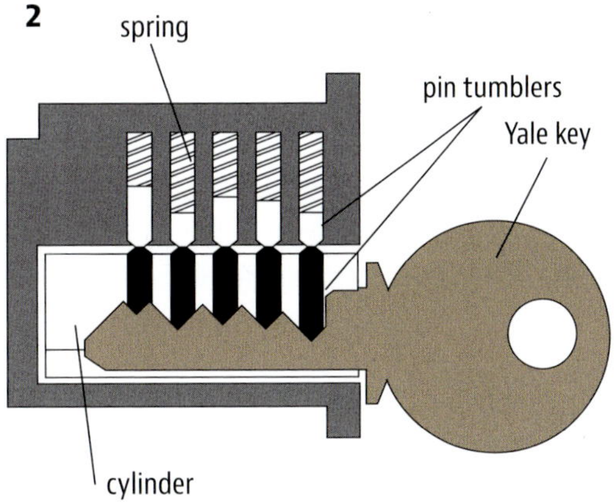

3

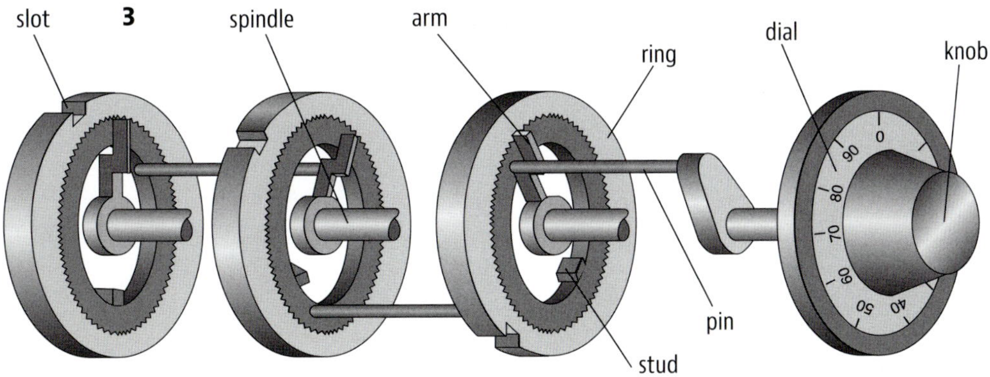

Hotels often use card keys, which contain a small microchip. The chip is programmed to open the lock of one specific room. If a guest loses their card, another card can be programmed.

Locks have been invented with many designs. Some resist explosives, becoming useless if tampered with; others have been designed to shoot, stab, or seize the hands of intruders. Locks have been made that can be locked with one key but only opened with another. The basic key types remain the warded, the lever, and the pin-tumbler, although many variations have been made, often combining the best features of each.

In buildings with many different locks, such as offices, schools, and prisons, it can be very useful to have a "skeleton key" that can open all the doors in the building. This master key can be shaped to avoid the warding of all the locks, or there may be two keyholes in the lock, one for the normal key, one for the master key. Alternatively, the locks can be fitted with two sets of tumblers and levers or, in the case of Yale locks, two concentric cylinders—one operating inside the other.

Today, like most other forms of technology, locks have become computerized. The code is no longer physically encrypted on a key but is magnetically or digitally encoded on a card. Some locks can recognize an individual's retina

RESTRAINING DEVICES

Handcuff locks have a row of teeth cut into one side of an open metal loop that can be pushed past a spring-loaded catch until the cuff fits tightly around the wrist. The catch prevents the cuff from being loosened unless the correct key is inserted. Disposable handcuffs, which do not include locks, are also available. Made from strong nylon, they can only be removed by being cut off. These lightweight restraints are particularly useful to police in crowd situations such as riots.

(the inner surface of the eye) or their fingerprints. One of the ultimate locks, used on the U.S. nuclear missile launch system, can only be activated by two operators putting different keys into keyholes at opposite ends of the wide control console and turning them at precisely the same moment. Even if one of the operators was to overpower the other, they would still be unable to launch the missiles. The operator would not be able to reach far enough to turn both keys at the same time. The race between people who devise locks and people who try to find ways of breaking through locks is a fast-moving area in the history of technology.

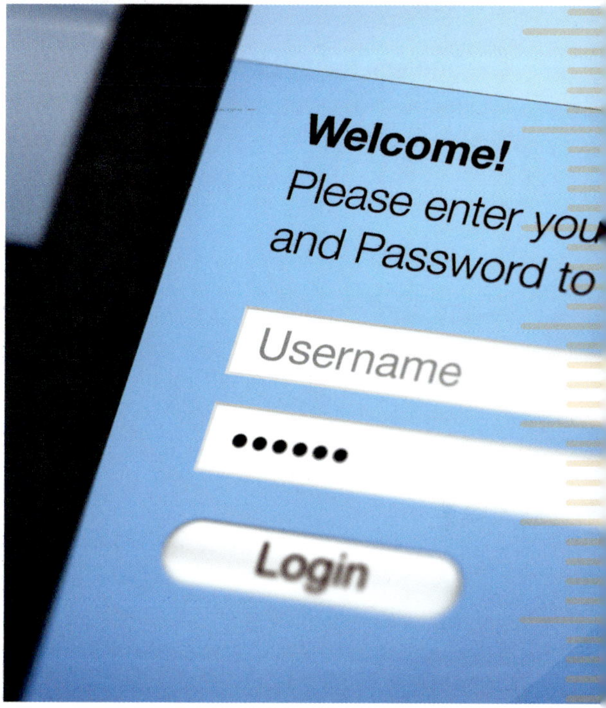

Computer security keeps unapproved people locked out of the system by asking for a password. The password system is not new. It was used by Roman guards, who challenged every visitor for that day's "watchword."

AI AND THE UNPICKABLE LOCK?

Artificial intelligence (AI) has made locks even more secure. It is possible to identify an individual without any risk of error by measuring particular biological features of the human body—a technique called biometrics.

Biometrics can include fingerprints, and iris or voice recognition. One reliable biometric lock works by scanning the retina. Each person has a unique pattern of blood vessels on their retina. The biometric lock uses artificial intelligence (AI) to recognize the retinas of people who are allowed to enter. There are no physical keys that could fall into the wrong hands, making this one of the most secure forms of lock.

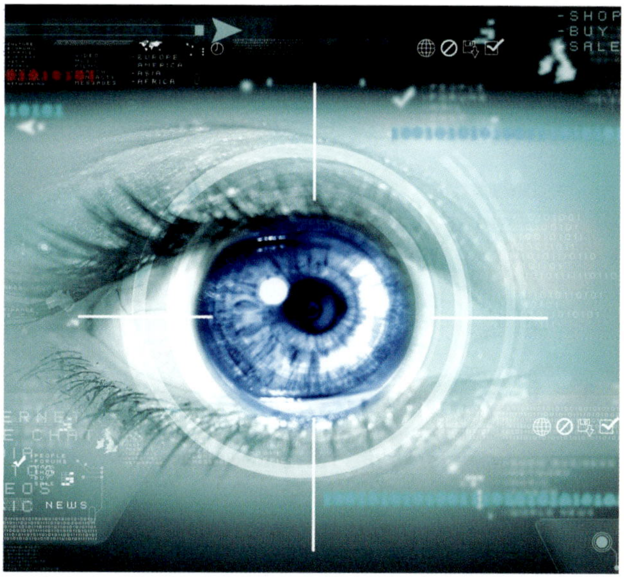

A biometric retina scan takes a picture of the retina and looks for a match among master copies stored in its memory.

TIMELINE

400,000 BCE The first evidence of wooden spears found in Europe, used by Neanderthals.

40,000 BCE The atlatl is invented in North Africa. It throws a dart.

23,000 BCE Throwing sticks are used in Europe and Africa, similar to the boomerangs of Australian Aboriginals, although most do not come back to the thrower.

20,000 BCE The first arrowheads date from this time, although no bows of this age are known.

c.8000 BCE In Europe, flint mines are being worked to make cutting blades.

3500 BCE Bronze is in use. Wheels are added to sleds to make the first carts.

3,000 BCE The composite bow, a small but powerful weapon, is invented in northern Asia.

2000 BCE The first lock is invented in Iraq.

1500 BCE Iron smelting is first carried out on a large scale. The Egyptians make body armor by attaching small pieces of bronze to leather garments.

500 BCE Trebuchet catapults are developed in China. The ballista crossbow device is invented in Greece around the same time.

1000 CE First known formula for gunpowder appears in a Chinese manuscript.

1100s The first true sailing warship, the cog, is built in northern Europe.

1400s Metal guns are used for the first time, originally in China and India, then in Europe.

1415 The longbow revolutionizes warfare in Europe.

1501 Gunport invented for firing cannons from lower decks of ships.

1771 First military submarine, the *American Turtle,* puts to sea.

1805 Congreve rockets used in naval bombardments of harbors.

1813 *Demologus*, the first steam-powered warship is launched.

1835 Samuel Colt patents the Colt revolver.

1836 Francis Smith and John Ericsson come up with a successful screw-propeller design.

1848 The breech-loading rifle, invented by Johann Dreyse, comes into service.

1860 HMS *Warrior*, the first iron warship, is built in Britain.

1860s The first Yale locks are manufactured.

1862 First engagement between ironclad ships takes place at Battle of Hampton Roads.

1875 Alfred Nobel invents dynamite.

1884 Hiram Maxim develops the first fully automatic machine gun.

1902 Trinitrotoluene (TNT) is invented in Germany.

1918 First aircraft carrier, HMS *Argus*, is launched.

1939 The first jet-powered airplane, the Heinkel He-178, is flown. The first practical helicopter is flown in the United States by engineer Igor Sikorsky.

1942 Enrico Fermi sets up the first self-sustaining nuclear chain reaction.

1943 The Colossus computers are developed in England to crack German military codes.

1944 The V-1 flying bomb, or "doodlebug," becomes the first cruise missile.

1945 The first atomic bombs are dropped by the United States on the Japanese cities of Hiroshima and Nagasaki.

1952 The first hydrogen bomb is tested by the United States.

1953 The United States develops a surface-to-air missile (SAM) by this time.

1955 USS *Nautilus*, the first nuclear-powered submarine, is launched.

2002 Lasers used to destroy artillery shells.

2007 Estonia suffers a cyberattack from Russia, turning off many of the country's Internet and telecommunication services.

2008 The first plasma weapon is tested, firing balls of super-hot gas at targets.

2014 Laser weapons are used for the first time on board a ship.

2020 A pilotless U.S. MQ-9 Reaper drone attacks and kills Iranian General Qasem Soleimani. The drone is controlled from a base thousands of miles away from its target.

2022 Ukrainian troops use drones to attack Russian tanks invading their country.

GLOSSARY

American Revolution (1775-1783)
A war fought between Great Britain and the 13 British colonies in North America. The first submarine was used in this war.

ancient Greece A civilization that existed on the mainland and islands of modern-day Greece and Turkey between 2000 and 300 BCE.

artillery Large guns designed to fire heavy projectiles such as explosive shells. Artillery pieces can be mobile, stationary, or mounted on ships or airplanes and are often capable of destroying targets many miles away. In ancient Greece and Rome, the word artillery described machines that propelled stones or other projectiles using the forces of torsion or tension.

atom The smallest unit of a chemical element that can take part in a chemical reaction. Atoms are composed of a central nucleus made up of protons and neutrons surrounded by shells of electrons.

bacteria Single-celled microorganisms that are present almost everywhere on Earth. Bacteria can only be seen through a microscope, and most are spherical, rodlike, or spiral in shape. Some bacteria are beneficial to people, helping us digest our food and playing an important part in the preparation of some foods. Others are responsible for serious diseases such as cholera.

American Civil War (1861-1865)
A war fought between the Northern (Union) and the Southern (Confederate) states. One of the major disputes was over slavery, which the South supported and the North wanted to stop. The first battle between ironclad naval vessels was fought during this war.

Cold War An extended period of hostility that ran from 1945 to the start of the 1990s. A group of communist states, led by the Soviet Union (a now dissolved communist empire centered on Russia), formed one side, and they were opposed by a group of noncommunist nations, dominated by the United States. Although conflict never broke out, there was a constant threat of war, and both sides produced vast numbers of nuclear and conventional weapons in case of attack. The Cold War ended with the collapse of many communist governments in the late 1980s and early 1990s.

digital A measuring system, such as binary code, that can store information as a number of discrete values. A binary computer is an example of a digital device since it uses only two values—one and zero—to store data.

frigate A small, fast ship used in sea warfare.

gamma rays High-energy electromagnetic radiation emitted after nuclear reactions.

Holocaust The systematic murder of six million Jews and millions of gypsies, Poles, and Russians by the Nazis during World War II. Victims of the Holocaust were executed, starved, or worked to death, often at specially constructed concentration camps.

Industrial Revolution A great change in social and economic organization brought about by the replacement of hand tools by machines and power tools, and the development of large-scale industrial production methods. The Industrial Revolution started in England around 1760 and spread to the rest of Europe and the United States. Industrial processes transformed all subsequent weapons and warfare.

infrared A type of electromagnetic radiation. Infrared rays have a strong heating effect and are emitted by hot objects.

Korean War (1950-1953) A war in which a U.S.-dominated United Nations coalition came to the aid of South Korea during an invasion by North Korea, which was aided by the Soviet Union and communist China. The war was ultimately indecisive, but was one of the key military confrontations of the Cold War.

medieval Of or from the Middle Ages.

metallurgist A person who studies the science of metals, particularly the extraction and processing of metal ores.

Middle Ages A period of European history from around 500 to c.1450 CE.

Nazi The name of a political party or its members that, led by Adolf Hitler, ruled Germany from 1933 to 1945. The Nazis suppressed all opposition and built up Germany's military strength. They were the main protagonists of World War II.

neutron A component of an atom's nucleus. Neutrons do not have an electric charge.

orbit The path, shaped like a circle or an ellipse, that an object in space takes around another object.

radiation The act of giving off radioactive particles, heat, or electromagnetic waves.

radio waves A type of electromagnetic wave with a wavelength between 1 mm and 1,000 m. Radio waves are used by people to transmit information over long distances.

radioactivity The disintegration of atomic nuclei accompanied by giving off particles or electromagnetic waves.

Rome The ancient civilization that began in the Italian city of Rome around 700 BCE and had established a vast empire around the Mediterranean by 200 CE. The Romans are noted for being the first to bring law and order to Europe and for their great works of engineering.

satellite A natural or artificial object in orbit around a star or planet.

Soviet Of or from the USSR, a communist state that existed from 1923 to 1989 and included present-day Russia.

telecommunications Sending messages over a distance, usually involving electrical signals or electromagnetic waves.

Vietnam War (1957-1975) A conflict between communist North Vietnam (supported by the Soviet Union and China) and noncommunist South Vietnam (supported by the United States). American soldiers were actively involved in the war from around 1965 to 1973, after which hostile public opinion in the United States forced their withdrawal. North Vietnam conquered the South in 1975.

virus A tiny, disease-causing particle that consists of protein combined with genetic material (DNA or RNA—ribonucleic acid). Viruses are only capable of replicating inside living cells, and for this reason many scientists do not consider them to be living organisms. Other scientists, however, consider them to be a type of microorganism.

World War I (1914-1918) A war fought mainly in Europe between the Central Powers—Germany, the Austro-Hungarian Empire (present-day Austria and Hungary), and the Ottoman Empire (now Turkey)—and the Allies: France, the British Empire, Russia, and the United States. The Allies eventually won the conflict, but millions of soldiers and civilians on both sides lost their lives. Air, submarine and tank warfare first became widespread during this war

World War II (1939-1945) The most destructive conflict in history, fought mainly in Europe, East Asia, and North Africa. The Axis powers (Germany, Austria, Japan, and Italy) were opposed by the Allies (Britain, the United States, France, and the Soviet Union). Germany surrendered in April 1945, but Japan fought on until August, when atomic weapons dropped by U.S. aircraft destroyed the Japanese cities of Hiroshima and Nagasaki.

FURTHER RESEARCH

Books

Fire and Stone: The Science of Fortress Warfare 1660-1860 by Christopher Duffy. New York: Helion and Company, 2023.

Warfare Through the Ages: From Bows to Bombs by Michael and Mary Woods. New York: Twenty-First Century Books, 2024.

Working With Tech in the Military (Technology in the Workplace) by Jeff Mapua. New York: Rosen Young Adult, 2020.

Web Sites

Ancient Greek Warfare
worldhistory.org/Greek_Warfare/

Technology of War
science.howstuffworks.com/fcs.htm

Unmanned Aerial Vehicles (UAVs)
britannica.com/technology/military-aircraft/Unmanned-aerial-vehicles-UAVs

INDEX